Monographs on Endocrinology

Volume 32

John W. Everett

Neurobiology of Reproduction in the Female Rat

A Fifty-Year Perspective

With 73 Illustrations

Springer-Verlag
Berlin Heidelberg New York
London Paris Tokyo Hong Kong

John W. Everett, Ph.D.
Department of Neurobiology
Duke University Medical Center
Durham, NC 27710, USA

ISBN-13:978-3-642-83799-9 e-ISBN-13:978-3-642-83797-5
DOI: 10.1007/978-3-642-83797-5

Library of Congress Cataloging in Publication Data.
Everett, John W.,
Neurobiology of reproduction in the female rat / John
W. Everett.
(Monographs on endocrinology ; v. 32)
Includes bibliographical references.
ISBN-13:978-3-642-83799-9 (U.S.)
1. Reproduction. 2. Rats--Physiology. 3. Neuroendocrinology.
I. Title. II. Series.
[DNLM: 1. Neurobiology. 2. Rats, Inbred Strains--physiology.
3. Reproduction. W1 M057 v. 32 / QY 60.R6 E93n]
QP259.E94 1989

Typesetting: International Typesetters, Inc., Palanan, Makati; Philippines

2127/3020-543210 – Printed on acid-free Paper

To Marian

Preface

It has been my privilege and pleasure during the past half century to participate in the unfolding of present-day concepts of the mammalian female reproductive cycles. When the studies recorded here began in the late 1930s it was already established that cyclic ovarian function is governed by gonadotropic secretions from the anterior pituitary gland, the "conductor of the endrocrine orchestra," and that in turn this activity is importantly dependent in some way upon secretion of estrogens and progesterone by the ovaries. Although a role of the nervous system was recognized for the reflex-like induction of ovulation in rabbits and cats and the induction of pseudopregnancy in rats and mice, and although there was even some evidence of neural participation in ovulation in rats, a major central neural role in the female cycle of most species was not apparent. Gonadotropic fractions of pituitary extracts having distinct follicle-stimulating and luteinizing activities in test animals had been obtained, and these respective effects had been fairly well characterized. Prolactin was well known for its lactogenic activity, but its luteotropic role in rats and mice had yet to be revealed. The molecular structure of the several estrogens and progesterone was known, and they were readily available as synthetic products. The broad concept of ovarian-pituitary reciprocity appeared to be an acceptable explanation of the female cycle, with the ovary in control through the rhythmic rise and fall in secretion of follicular estrogen. George Corner in his book *The Hormones in Human Reproduction* (Princeton University Press, 1942) would include a chapter entitled *The Ovary as Timepiece.*

There were known to be modifying influences of the environment on reproduction, however, which in themselves implied some participation of the nervous system; seasonal changes were obvious in many species. In rats, it was known that exposure of cycling females to continuous illumination would induce a persistent state of vaginal cornification, failure of ovulation and the presence of polyfollicular ovaries lacking corpora lutea. The existence of a "sex center" in the rat hypothalamus had been proposed, and the importance of the pituitary stalk for ovulation had been demonstrated in both rabbits and rats. However, the special importance of the hypophysial portal vasculature for anterior pituitary regulation was speculative at best.

It was fortuitous that after coming to Duke University, for continuing my studies of the rat placenta I felt it necessary to use animals of the same stock that I had used in New Haven; I chose to use close inbreeding by brother-sister matings. Probably as the result of the mixed heredity of the breeding stock the inbreeding soon brought out the tendency toward the occurrence of spontaneous persistent vaginal estrus in the early life of nonbreeding females and toward a special sensitivity to environmental illumination. These striking features begged for investigation, diverted me from the placenta and led to the long series of investigations that are the subject matter of this monograph.

August, 1989 John W. Everett

Contents

Introduction 1

Vaginal Cytology and Ovarian Status 2

Strains of Rats 8

Interaction of Environmental Lighting, Age, and Genetic Background Influencing the Estrous Cycle 9

Spontaneous Persistent Estrus (SPE) 9
Influence of Seasonal Variation of Daily Illumination 11
Persistent Estrus Induced by Continuous Illumination (LLPE) 14

Steroid Regulation of the Ovarian Cycle 17

Restoration and Maintenance of Ovulatory Cycles by Progesterone Treatment of Persistent-Estrous Rats 17
Advancement of Ovulation by Progesterone in Normally Cycling Rats 22
Corpora Lutea of Cyclic Rats as Possible Sources of Progesterone 24
Induction of Ovulation by Estrogen 33
Ovulation Advanced by Estrogen in Cyclic Rats 35
Strain Differences in the Ovulation Response to Estrogen 37

Timing the Preovulatory Surge of Gonadotropin Secretion 38

Blockade of Estrogen-Induced Ovulation 39
Blockade of Progesterone-Induced Ovulation 43
Blockade of Spontaneous Ovulation 46
Pharmacologic Intervention Compared with Hypophysectomy During the 'Critical Period' 47
Effect of the Lighting Rhythm on the 'Critical Period' 51
Circadian Periodicity in Mechanisms Governing the Ovulatory Surge of Gonadotropins 52
The Empirical Nature of the 'Critical Period' 56

Central Neural Control of Spontaneous Ovulation: The Rat Model 57

Electrochemical Stimulation (ECS) 57
Electrical Stimulation (ELS) 62
The Septal-Preoptic-Tuberal System (SPTS) and the Ovulatory LH Surge 63
Non-Specific Stimulation of the Tuber by Destructive Lesions 66
Dose-Response Relationship of Stimulation to LH Release and Ovulation 69
Electrical Stimulation of MPOA versus Basal Tuber 73
The Comparative Increments of Circulating LH with Increasing Time of Electrical Stimulation of the Medial Preoptic Area or the Arcuate Nucleus-Median Eminence 80
Similarity of LH Surges by MPOA Stimulation in Rats Blocked with Various Drugs 83
Stimulation of the Preoptic-Tuberal System During Persistent Estrus......... 87
Overview of the Stimulation Studies 90

Pseudopregnancy: Controls for Prolactin Secretion 92

Removal of the AP from Direct Influence of the Hypothalamus Favors Secretion of Prolactin 93
Pseudopregnancy Induced by Damage to the AP, Stalk-Section, or Damage to the Basal Hypothalamus 98
Re-transplantation of AP Grafts from Kidney to Brain 99
Delayed Pseudopregnancy: An Approach to Mechanisms in Prolactin Control 106

Summary 115

References 119

Acknowledgments 130

Subject Index 131

Introduction

Embryologists searching for early stages of mammalian development were responsible for the first basic knowledge of circumstances attending ovulation. The relationship was recognized in a general way between behavioral estrus, on the one hand, and ovulation, corpus-luteum formation and pregnancy on the other. Aside from cats and rabbits, in which ovulation is provoked by copulation, spontaneous ovulation was evident in familiar species. Rhythmic changes were known to occur in the reproduction tracts of spontaneous ovulators, but the close relationship between such changes and the time of ovulation remained a mystery until the vaginal smear technique was devised for the guinea pig (Stockard and Papanicolaou 1917), the rat (Long and Evans 1922) and mouse (Allen 1922). The following excerpt from the classical monograph of Long and Evans is noteworthy:

"By the fortunate discovery that in the guinea pig [cyclic mucosal changes in the reproductive tract] are accompanied by dehiscence of epithelial cells so that at times the vagina has a characteristic cell content, it has been possible for Stockard and Papanicolaou to show us that we may discover with ease in the living animal the exact occurrence and progress of these cycles. When it has been proved, as Stockard and Papanicolaou have done for the guinea pig, and as we have been able to do with exactitude for the rat, that these cycles are correlated with the rhythmic discharge of ova from the ovary, it will be seen that we now have in our hands for the first time an accurate method for the detection of ovarian function in experimental animals. This fact promises important consequences, for it enables us to investigate disturbances of ovarian function which may be experimentally produced."

In the laboratory rat, Long and Evans demonstrated for the first time that vaginal cycles of 4 or 5 days duration are the rule; they were "inclined to regard [the 4-day cycle] as the true, normal cycle." Stages of the typical cycle were correlated histologically with events in the ovary, uterus and mammary glands. Normal and abnormal modifications could be detected through the daily variations in vaginal cytology. Thus, either infertile copulation or mechanical stimulation of the cervix prolonged the diestrous (leukocytic) phase of the cycle to 10–12 days or longer, a pseudopregnancy marked by suspension of ovulation and by the enlargement of the current set of corpora lutea as in pregnancy. Means were devised for differentiating the successive sets of corpora lutea. Experimental production of deciduomata was accomplished during pseudopregnancy, but not during the short cycle. Ovarian transplants could restore cyclic luteinization to ovariectomized hosts, and such transplants presented the usual pseudopregnancy responses to cervical stimulation. Ovaries from 21-day old donors, when transplanted to ovariectomized adult females, rapidly matured to functional status which could be recognized by the vaginal smears.

Vaginal Cytology and Ovarian Status

Beginning with studies reported by Everett (1939), more than one million vaginal smears from over 16,000 rats have been prepared and evaluated in the author's laboratory during approximately 5 decades. It is appropriate to consider criteria for evaluating the smears and their relationship to ovarian function. Our criteria are based on the five stages of the estrous cycle of rats described by Long and Evans (1922) and presented here in Table 1 essentially as originally summarized. It is important to recognize that each stage represents a stop-action in a steadily moving process. The stages are to be regarded as indirect and somewhat delayed measures of ovarian hormone output. Thus, in general terms, the rise of estrogen in late

Table 1. Stages of the rat estrous cycle correlated with status of the vaginal mucosa, uterus, and ovaries. Modified from Long & Evans (1922)

Stage	Vagina	Uterus	Ovary & Oviduct
I Early proestrus	Mucosa dry. Lips somewhat swollen. Smear: epithelial cells only.	Becoming distended.	Large follicles, not hyperemic.
II Late proestrus	Mucosa dry. Lips swollen. Smear: cornified cells only.	Maximal distension.	Large follicles, hyperemic late.
III Estrus	Smear: Masses of cornified cells	Fluid discharged.	Ovulation, fresh corpora lutea. Ova in oviduct.
IV Metestrus (brief)	Mucosa becoming moist. Smear: mixed leukocytes, cornified and epithelial cells.	Contracted	Corpora lutea organizing.
V Diestrus	Mucosa thin, moist. Smear: leukocytes, scattered epithelial cells.	Slender	Corpora lutea well formed.

diestrus results in proliferation of the vaginal epithelium to give the dry, velvety appearance characterizing proestrus (stages I and II).

By the morning after ovulation (stage III) the proliferative effect of estrogen has reached and passed its peak. The surface epithelium, now heavily keratinized, begins to slough away in white masses. Later that day a vaginal smear discloses nucleated cells from deeper levels of the epithelium and even a moderate number of leukocytes mixed with the cornified cells. This is the beginning of the transient vaginal stage IV, characterized at its peak by a pronounced invasion of leukocytes among residual clusters of cornified cells. The transition is completed within a few hours, after which the vaginal smears are predominantly leukocytic for two or three days (stage V). As diestrus advances, the leukocyte population diminishes, and they become scarce with the approach of the next proestrus.

In our routine, the smears have been made by lavage with physiological saline, thinly spread on a chemically clean glass slide for rapid drying. If the fluid in the pipet remains clear after the first lavage, the vagina is repeatedly flushed; otherwise the first wash is sufficient. If the fluid is exceedingly milky, pains are taken to leave a minimum film on the slide. To enable interpretation of the record for each rat, her smears for each day of the week are placed in fixed order on a single slide. At the end of the week the slides are stained in 1% aqueous toluidine blue O after fixation in 95% ethanol and removal of salt by washing in de-ionized water. The dye solution must be neutral or slightly alkaline, else the cornified cells remain colorless. Nuclei and mucus are stained metachromatically pink, while the cytoplasm is in various shades of blue, very dark in the case of small epithelial cells, on the one hand, to pale blue in the squamous cells whether or not cornified.

There is only a rough correspondence between *stages* (Table 2) of the vaginal cytology (identified as I, II, III, IV, and V) and the successive *days* of the estrous cycle. We identify the days as proestrus, estrus, diestrus-1, diestrus-2 or diestrus-3 (Table 3). In this usage, proestrus (P) is that day when the spontaneous ovulatory surge of gonadotropin is released from the pituitary gland, hence the day preceding ovulation. Estrus (E) is the day on which ovulation occurs in the early morning. Diestrus (D-1, etc.) constitutes the interval in days between E and the next P, whatever their number (Fig. 1). Since the transition from the heavily cornified epithelium of estrus to the thin, moist epithelium of diestrus is abrupt, we omit the term "metestrus" in designating days of the cycle, using it only to refer to the stage IV of vaginal cytology.

Evaluation of the vaginal smears prepared daily for two weeks or more will give satisfactory evidence of regularity of cycle length, for predictability of the day when the ovulatory surge of gonadotropin occurs and, hence, for the approximate time of ovulation. However, to make such evaluations one must recognize variations in the composition of the smears that may be encountered among strains of rats, from cycle to cycle in the same individual, and especially in cycles of different length (Tables 2 and 3). In addition, there are variations due to technique and to time of day. In agreement with Long and Evans (1922) we have found that the 4-day cycle length is the most frequent and we regard it as the standard length of cycle in adult rats maintained under laboratory conditions. We consider the 5-day cycle to be a normal variant in which for some reason the secretion of estrogen in late diestrus is somewhat

Table 2. Changing composition of the vaginal smear throughout the rat estrous cycle, with codes for the sequential stages

Code	V_{ec}	V_{EC}	I_A	I_B	I_C	II	III	III^e_n	III^E_{N1}	IV	IV_{e1}	V
Epithelial cells												
Small, basophilic, diffuse	++	++	+++	±	±	–	–	–	±	±	++	++
clusters	–	–	++	±	±	–	–	–	–	–	–	–
Medium, pale, nucleated	–	–	+	+	±	–	–	±	+	+	+	±
not nucleated												
round	–	–	+	+++	+	–	–	±	+	+	+	±
angular	–	–	+	+++	+++	+++	±	–	–	–	–	±
Large, cornified, not clumped	+	+	+	++	+++	+++	+++	+++	+++	+++	+	+
clumped	–	–	–	±to++	±to++	+to++	+++	+++	+++	+++	++	±
Large, squames, nucleated (Schorr cells)	±	±	±	±	±	±	–	+	++	++	++	±
Leukocytes	++	+ (trace)	±	–	–	–	–	–	+	++	+++	++
Bacteria	–	–	±	+	+++	+++	+to–	+to–	–	–	–	–
Density	thin	very thin	increasingly milky .... /			.... heavy, coarse			.... /	....	very milky	

Table 3. Vaginal smear patterns and their variants commonly presented by cycling rats in the 1000 h-1500 h period[a]

4-day cycle

Diestrus day 1 (D-1)	Diestrus day 2 (D-2)	Proestrus (P)	Estrus[b] (E)
IV	V_{ec}	V_{EC}	II
IV_{el}		I_A	III
V		I_B	III

5-day cycle[c]

Diestrus day 1 (D-1)	Diestrus day 2 (D-2)	Diestrus day 3 (D-3)	Proestrus (P)	Estrus[b] (E)
IV_{el}	V_{ec}	V_{ec}	I_A	III
V		^{V}EC	I_B	III^{e}_{n}
		I_A	I_C	$III^{E}_{N}l$
			II	IV
			III	

[a] The days of the cycle are only approximately correlated with the vaginal stages. "Metestrus", comprising vaginal stages III^{e}_{n}, $III^{E}_{N}l$, IV, and IVel, proceeds rapidly and is largely complete by the morning of D-1. Thus, this term is omitted.

[b] Heat and ovulation occur during the night before estrus, the LH surge occurring on the preceding afternoon.

[c] In the 5-day cycle, the relatively great variability of the smear pattern on D-3, P and E may lead to faulty interpretation.

retarded (Schwartz 1969) and the ovulatory gonadotropin surge is delayed by a quantum interval of 24 hours. Cycles extending for 6 days or longer are comparatively infrequent and likely to appear in animals that display repeated irregularity of cycle length.

Figure 1 illustrates typical vaginal smears characterizing the several days of 4-day and 5-day cycles, respectively, as they appear during the late morning hours (see also Tables 2 and 3). While the smears on D-1 and D-2 are alike in the two kinds of cycle, significant differences appear thereafter. On the third day, (P in the 4-day cycle, D-3 in the 5-day cycle) the range of variability is especially pronounced. At the one extreme, almost exclusively in the 5-day cycle, leukocytes may remain fairly abundant. Yet, in other cases, D-3 of the 5-day cycle may present cell populations common to early proestrus of the 4-day cycle (I_A to I_B). The progression of change in the cell population in 5-day rats is such that, if the morning smear on D-3 is in stage I_A, it will have advanced next morning to I_C, II or even III. Although full cornification may then seem to have been reached, examination of the vagina with a speculum will disclose the smooth, dull pink epithelium characteristic of proestrus; the orifice will be turgid and hyperemic. This is the day before ovulation. The ovulatory surge of gonadotropin occurs that afternoon and the female becomes sexually receptive that

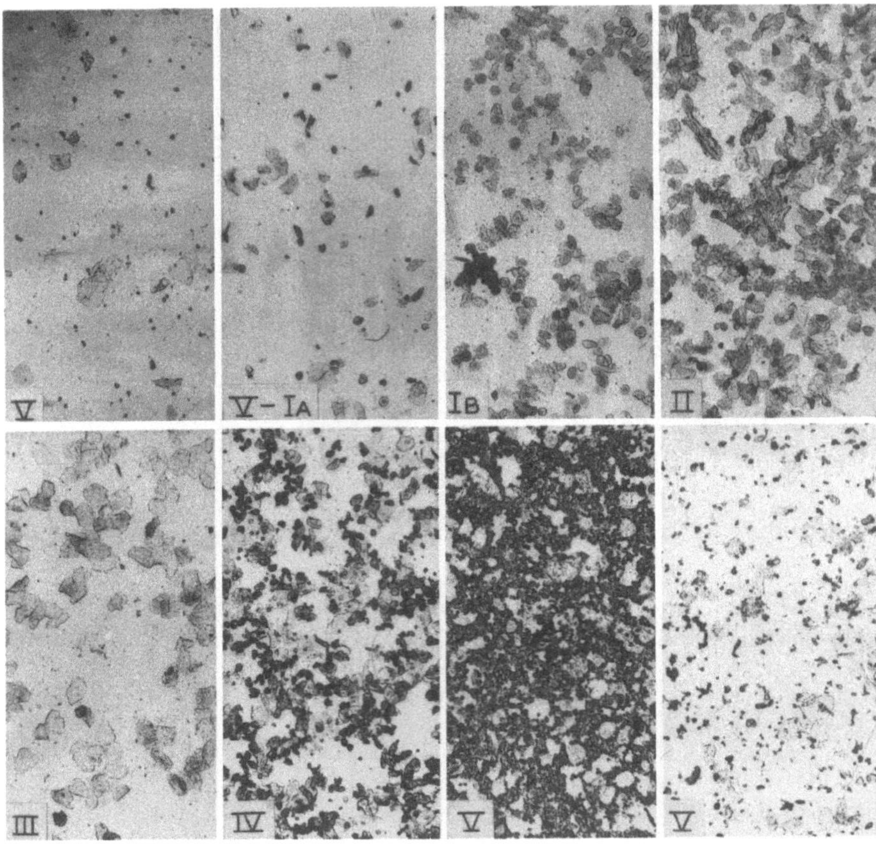

Fig. 1. Representative vaginal smears in the rat estrous cycle. The respective codes are detailed in Tables 2 and 3

night. Occasionally, especially in some strains of rats and more frequently in the 5-day cycle, receptivity may occur during the night preceding proestrus (Aron et al. 1961, 1966). Copulation then may lead to reflexive release of the ovulatory gonadotropin and ovulation.

On the day of vaginal estrus, i.e. the normal day of ovulation, there is again a rather wide variation in the cell population of the morning vaginal smear. In the 4-day cycle, it is nearly always in stage III at that time, heavily cornified, with desquamation of white, cheesy masses of epithelium. In the 5-day cycle, considerable numbers of small nucleated squamous (Schoor) cells and even a few leukocytes are often included. These elements herald stage IV, which may be well advanced later that day.

In a sequence of vaginal cycles, the final day of vaginal cornification will usually appear in the record at regular 4-day or 5-day intervals. Thus, the sequence III, V,

V, I, III, III, V, V, V, I, III, V represents two 5-day cycles and III, V, V, I, III, V, V, I, III, III V represents a 4-day and 5-day cycle in sequence. Persistence of vaginal cornification beyond two days or prolongation of diestrus beyond 3 days are both considered abnormal. The former indicates an incipient state of *persistent estrus*. Even the slow onset of proestrus in the 5-day cycle suggests this and spontaneous persistent estrus is often preceded by a number of 5-day cycles and/or by pronounced irregularity of cycle length.

Once persistent estrus has become established, the vaginal smears do not necessarily remain fully cornified day after day. Commonly there is a pseudocyclicity such as that occurring in ovariectomized rats receiving a steady supply of extrinsic estrogen (Del Castillo and Calatroni 1930; Del Castillo and Di Paola 1942; Hartman 1944; Zuckerman 1938). The population of cornified cells is accompanied from time to time by considerable numbers of nucleated epithelial cells and leukocytes. However, there is no conspicuous sloughing like that in the post-ovulatory estrus of normal cycles. Although some authors have reported that cycles in "old" rats may be anovulatory (Wolfe, Burack and Wright 1940; van der Schoot 1976), one may question whether these cycles truly represent ovarian cyclicity.

Strains of Rats

Most studies in this laboratory have been carried out in three strains of rats: two locally inbred strains, designated DA and O-M, respectively, and the commercial CD strain from the Charles River Breeding Laboratories.

The DA strain was of mixed origin. Wistar albinos in J.S. Nicholas' colony at Yale University had been outcrossed several times with wild gray rats and once with the Long-Evans Berkeley hooded strain. Our stock came from two albino pairs from the Nicholas colony brought to Duke University in 1932. After close inbreeding by brother-sister matings instituted in 1936, fertility problems soon appeared. Among the peculiarities was the fact that no females were fertile unless initially bred by the age of 100 days. If mating before that age was successful, however, several litters could be expected throughout the first year, provided that the doe was promptly returned to the buck after each litter was weaned. With attention to this trait and with continual selection for the best breeding history in several preceding generations, the DA strain was maintained until 1954, when all young females were accidentally allowed to become too old for breeding. DA males remained fertile, however, and for several years we kept a hybrid stock descended from a DA male and an O-M female.

The O-M strain, originating from the Osborne-Mendel colony, was maintained for several years at Vanderbilt University, then at Albany Medical College (V-S strain; Wolfe, Bryan and Wright 1938) and later at Duke University. We began our *inbred* line in 1941. This vigorous stock served through more than 100 generations until 1972.

Rats of the CD strain were our chief experimental subjects since 1968. The CD strain was derived from Sprague-Dawley rats at the Charles River Breeding Laboratories (Wilmington, MA, USA) and bred there as a "caesarian-derived, barrier-sustained" randomly outbred stock. The strain has been designated officially by the acronym Crl:CD BR. The Charles River colonies are maintained under controlled illumination of 12 hours daily. We commonly purchased females in the weight range of 125–150 g, at about 40 days of age. After arrival they were allowed to adapt to the local environment at least two weeks, after which vaginal smear records were obtained for at least another two weeks before experimentation.

Interaction of Environmental Lighting, Age, and Genetic Background Influencing the Estrous Cycle

An outstanding characteristic of the DA strain and its fertility problem was the suspension of cyclic estrus early in adult life among segregated females and replacement of cycles by more or less continuous vaginal cornification. This persistent vaginal "estrus" when well established was associated with the presence in the ovaries of prominent vesicular follicles and the absence of corpora lutea. Cycling DA females also rapidly developed the condition when exposed to continuous illumination. In addition, there were certain seasonal influences relating to the length of daily illumination as detailed below.

Spontaneous Persistent Estrus (SPE)

Evans and Long (1921) were first to describe this phenomenon. Asdell and Crowell (1935) later reported that 71% of their rats presented prolonged vaginal estrus at some time in life. Their finding and that by Wolfe, Bryan and Wright (1938) indicated a relationship to aging. This was clearly shown by Wolfe, Burak and Wright (1940) who also noted a marked difference in the ages at which the condition appeared in two strains of rats, the Albany A-S strain and the Vanderbilt V-S strain derived from the Osborne-Mendel colony. The V-S strain, highly fertile, did not exhibit defects in the vaginal cycle and ovarian histology until well into the second year of life, much later than in the A-S strain. More recent reports describe SPE as a feature of beginning senescence in the female laboratory rat, appearing during the second year and often followed by a period of repetitive pseudopregnancies and eventual ovarian atrophy and vaginal anestrus (Aschheim 1961; Bloch and Flury 1959; Huang and Meites 1975). However, findings with the DA strain show that SPE can occur surprisingly early in life and that the age factor, whatever it may be, has a genetic basis.

It is instructive to compare the ages of first appearance of SPE in rats of differing breeds housed in the same quarters. In virgin DA females the condition commonly appeared during the fourth or fifth month of age, a time when full reproductive vigor should prevail (Everett 1939). As shown in Fig. 2, two breeding lines of that strain showed somewhat different ages of onset. Most DA_c females experienced SPE before they were 180 days old, 50% by 135 days. DA_f rats reached the 50% mark at about 210 days. By contrast, Wistar rats of comparable ages presented regular 4-day or 5-day cycles. The same was true of O-M rats: one group of 22 O-M rats, set aside and followed by regular vaginal smearing until about 500 days of age, rarely displayed any prolonged vaginal cornification before the end of the first year (Fig.

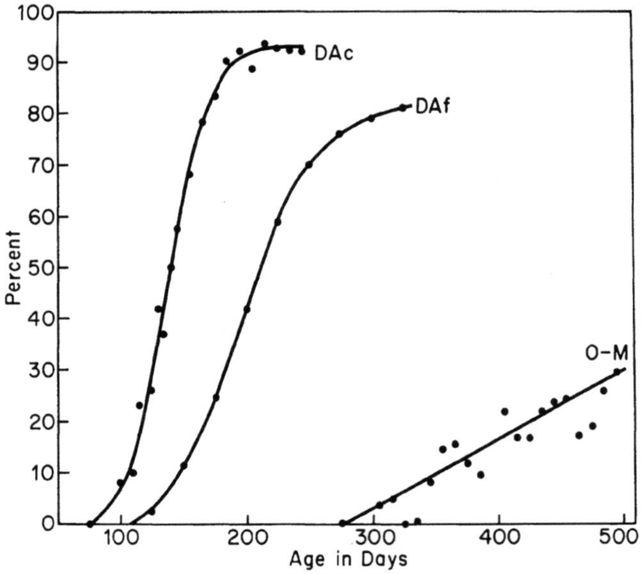

Fig. 2. Comparative ages of onset of spontaneous persistent estrus in two DA substrains and the O-M strain. Each point represents the percentage of rats showing prolonged estrous smears at a given age. (From Everett 1970)

2). Throughout 31 years of experience with the inbred O-M strain, we could expect regular cycling in animals under a year old. First and second generation hybrids between the O-M and DA_c strains showed onset of SPE at ages intermediate between those in the parent strains. In the CD strain, we have frequently encountered SPE after the ages of 5 to 6 months in animals that were by no means senescent in other respects. Thus, not only is the onset of SPE a function of age, but also a function of genetic background.

The pattern of early onset of the condition in the DA and CD strains bears a resemblance to the "delayed anovulatory syndrome" in rats neonatally treated with small doses of androgen (Arai 1971; Gorski 1968; Harland and Gorski 1978; Swanson and van der Werff ten Bosch 1964). Administration of 10 mg testosterone propionate to female rats on the fifth postnatal day typically results after puberty in a succession of normal short cycles for several weeks before the eventual onset of persistent estrus. While the resemblance may be no more than superficial, it is conceivable that the relatively early onset of SPE in certain adult rats may reflect some predisposing peculiarity in perinatal steroid secretions. Another important factor timing SPE is the environmental illumination, as shown below.

Influence of Seasonal Variation of Daily Illumination

When spontaneous persistent estrus and other defects of reproduction were first encountered in the DA strain the colony was housed in a room with undarkened windows. The length of daily illumination was thus determined by sunrise and sunset. During the short days from early December to early March, when fertility was comparatively poor among animals selected for breeding, segregated virgin females tended to present estrous cycles with prolonged diestrous phases or even to become anestrous. Oddly enough, females that had become persistent-estrous tended to return to cycling. Illustrative examples are charted in Fig. 3 and 4. With the lengthening daylight in spring, on the other hand, the (younger) cycling animals became more regular and SPE returned to the older rats that had temporarily resumed vaginal cycles. The critical day length for the shift seemed to be in the 10- to 11-hour range. The influence of both shortening and lengthening days was expressed in the vaginal smears after delays of several days or even weeks. For instance, in the first example the onset of anestrus was approximately 10 days after the day length dropped to 10 hours. Cycles began to return in February about 3 weeks after the day length had returned to 10 hours. Inasmuch as seasonal factors other than length of daylight *per se* could have influenced these results, it was necessary to examine the effect of light while the rats were under otherwise comparable conditions.

Two identical light-tight enclosures were placed side by side in the colony room. Controlled lighting in each was from a 25W incandescent bulb suspended about 20 inches above the floor of the nearest wire mesh cage. Ventilation by forced draft from small fans was designed so that air flow to the exhaust ducts passed close to the lamps.

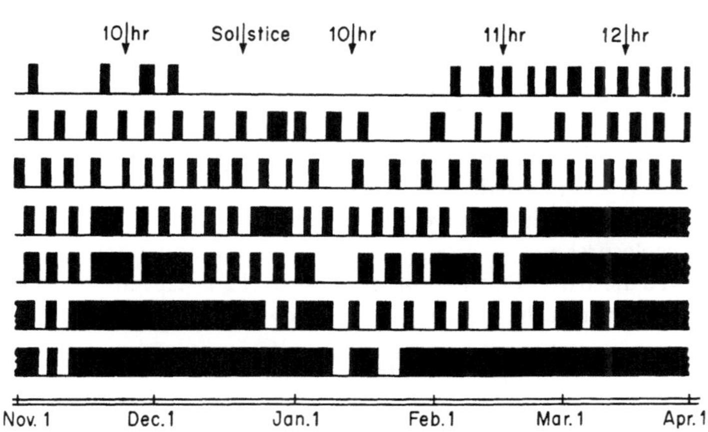

Fig. 3. Effects of seasonal changes in day length on estrous cycles in typical rats of the DA strain. Vaginal estrous stages (proestrus, estrus, metestrus) represented by black bars, diestrus by base lines. As day length shortened, cycling rats gradually tended to have prolonged diestrus, while rats already in persistent estrus tended first to regain cyclic function and then later to have prolonged diestrus. (From Everett 1970)

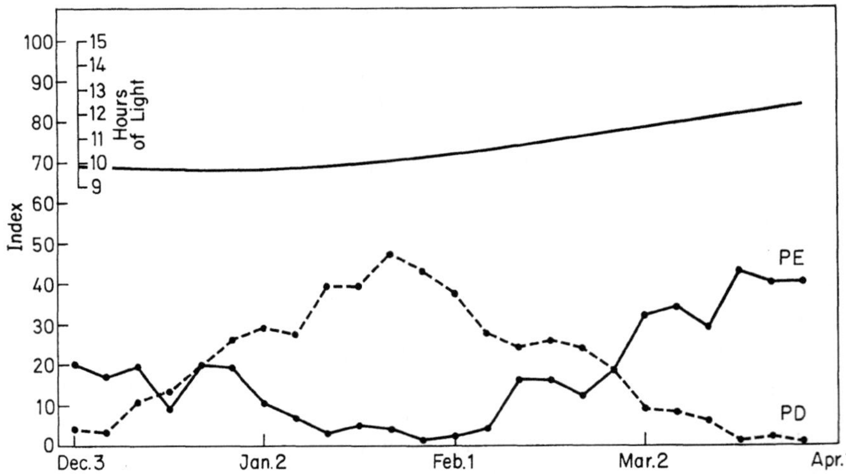

Fig. 4. Percentage incidence of prolonged estrus (PE) and prolonged diestrus (PD) in a group of DA rats exposed to natural day length. The "index" is the percentage of all rats whose current vaginal smears were examined. (From Everett 1970)

Measurement of the intensity of light at the cage floors gave values ranging from 29 to 75 lux, within the range observed in cages in the general colony. Air temperatures in the enclosures measured from day to day with maximum-minimum thermometers, only rarely exceeded room temperatures by more than 2°C.

Two experimental groups A and B, of 15 rats each, were selected (Figure 5), and a control group of 38 rats remained exposed to ordinary colony conditions. As much as possible, litter mates were represented among all three groups and the age ranges on January 1 were similar (102 to 185 days in A and B, 102 to 205 days in the controls). Beginning on February 1, when natural day length was about 10½ hours, groups A and B were placed in the enclosures, 3 or 4 rats to the cage. In enclosure A, the daily illumination was increased 15 minutes per day to an eventual maximum of 14½ hours. In enclosure B, day length was decreased first to 10 hours and 12 days later to 9½ hours. These respective conditions were maintained until March 29. Then the lighting for the two enclosures was reversed gradually, 15 minutes per day.

Indexes of prolonged diestrus (PD) and prolonged estrus (PE), respectively, were determined for each group in each 5-day period, as follows:

$$\text{PD Index} = \frac{N^{PD}}{N} \qquad \text{PE Index} = \frac{N^{PE}}{N}$$

where N = total number of rat days (no. of rats × 5).
N^{PD} = days of diestrus or anestrus greater than three per cycle.
N^{PE} = days of estrus greater than expectation.

Figure 5 shows the results. In group A the PD index was high during January, but may have begun a downward shift by the time the experimental conditions were imposed. After that the downward trend was marked. In contrast, the PD index in

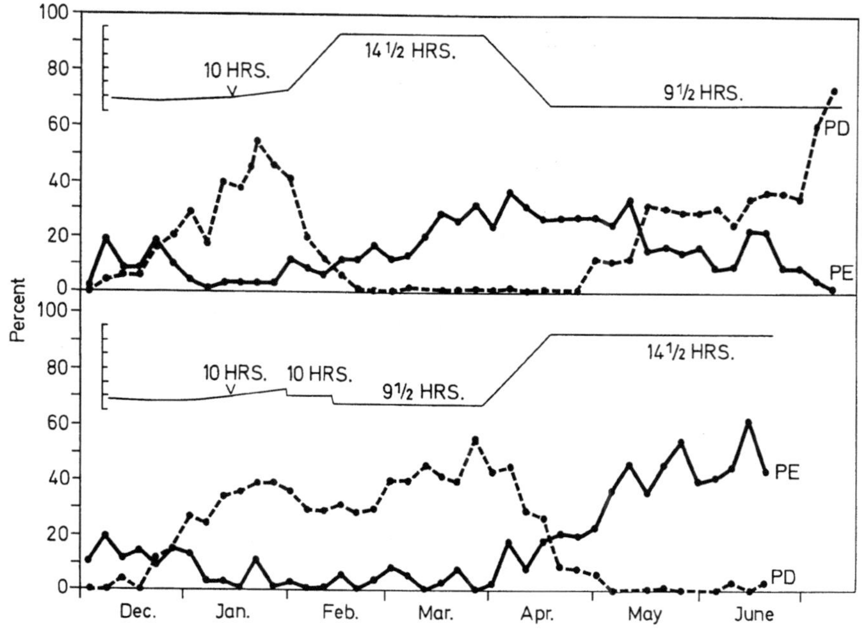

Fig. 5. Percentage incidence (index) of prolonged estrus (PE) and prolonged diestrus (PD) in two groups of DA rats exposed to controlled lighting. Length of daily lighting is shown by fine lines at top of each chart. (From Everett 1970)

group B was sustained and even somewhat increased throughout the two-month exposure to short days. Two rats remained anestrous throughout. As soon as the daily illumination increased, the PD index in group B fell as abruptly as that in group A had fallen two months earlier. Conversely, in group A the return to short days during April brought a pronounced rise in the PD index. The PE index, comparatively low in December and trending further downward in January, rose steadily in group A during February and March, much as in controls. In group B, by contrast, it remained low until daily lighting increased in April. The effect of short days on group A during April, May and June was to gradually lower the PE index.

The relatively more rapid climb of the PE index during long-day exposure in group B than in group A may be ascribed to the age difference, in view of the clear-cut effect of age on the incidence of SPE. Over all, the rates of change of the PE index and of the PD index under the experimental conditions were little different from the rates in controls for which day length changed far more slowly. The critical factor was evidently some particular duration of illumination (or darkness) and not the rate of increase or decrease. Above a critical ratio of light: darkness, regular cycles or persistent estrus prevailed, while below that ratio cyclic diestrus lengthened in the younger animals and persistent estrus gave way to cycles in the older rats.

To my knowledge, none of these seasonal effects has been reported in other strains of rats. Aschheim (1965), however, described restoration of short estrous cycles in senile SPE rats subjected to constant darkness. I did not examine the influence of darkness or short-day length in O-M rats. By the time they were brought into the colony, daily illumination had been standardized at 14 hours of artificial lighting throughout the year. The possible influence of short days was tested in 5 CD rats exhibiting SPE; when they were transferred to an enclosure lighted for only 8 hours per day, none showed any interruption of vaginal cornification during 8 weeks.

One effect of light, however, appears to be ubiquitous among all strains of rats: the induction of persistent estrus by exposure to continuous illumination.

Persistent Estrus Induced by Continuous Illumination (LLPE)

Hemmingsen and Krarup (1937) and Browman (1937) were first to record that continuous exposure of rats to light will bring about persistent vaginal cornification. As in SPE, once LLPE becomes established, corpora lutea are no longer present and the ovaries contain sets of large follicles, replaced from time to time as older follicles age. The phenomenon has become widely recognized as one of several means whereby persistent estrus can be induced experimentally and exploited.

The rapidity with which LLPE develops varies among strains of rats, among individuals of a given strain and, like other effects of light, seems to be governed in part by the animal's prepubertal experience with illumination (Hoffmann 1970, Takeo et al. 1975). Fiske (1941) reported that rats exposed from early life to continuous light continued to have corpora lutea for several months.

Extremely prompt induction of LLPE was discovered in DA rats when the lights in the colony room were unintentionally left burning during a holiday period. The effect was examined experimentally (Everett 1970) by placing two age groups of DA virgin females in the enclosures described above, with lights on continuously for 10½ days (Fig. 6). One group comprised 19 rats that were 68–76 days old at the beginning. In the other group of 19 rats the ages were 145 to 177 days. Most of the younger animals completed one short cycle after being placed in the light chamber, then a normal diestrus followed by prolonged vaginal estrus. One rat continued to show normal cycles, while 4 rats presented PE without the intervening cycle. In the older group, nearly all passed directly into PE; the intervening cycle appearing in only three subjects.

Two other strains were examined in similar tests: a locally raised Wistar strain and the O-M strain, newly acquired at that time. Among 9 Wistar rats 65 to 78 days old, only one began to show PE (after two cycles) during 10½ days of continuous lighting. Of 6 rats aged 107 to 111 days, only one began PE at the very end of the 10½-day test. Somewhat older rats, aged 141 days, gave a pronounced response in 3 of 4 cases, each completing the cycle in progress at the start.

O-M rats were tested in 4 age groups: 72 to 106 days, 152 to 153 days, 207 to 212 days, and 251 to 257 days of age at the beginning of exposure. The second and third groups are combined in Fig. 6. In the youngest group of 31 rats, not one presented PE during 10½ days. Eleven were left in the light for a total of 40½ days; only one of these showed prolonged estrus, two extra days of cornification in one cycle during the final week. Among nine 150-day old rats in the 10½ day test, three continued to

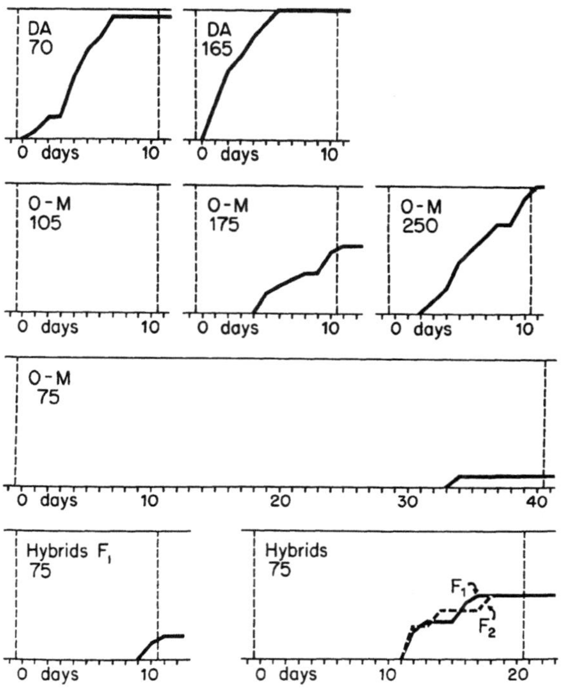

Fig. 6. Comparative rapidity of onset of prolonged vaginal estrus in DA_c rats, O-M rats and hybrids exposed to continuous illumination at different ages. In each graph 100% of the test group is represented by the top line and duration of the illumination is shown below. (From Everett 1970)

have normal cycles, while six presented PE after one or two short cycles. In the 210-day group, six of the ten rats continued to cycle normally, while four developed PE after one or two short cycles. Only the 250-day group presented a response in all animals: in one of the 10 rats at the first estrus, in the others after one or two cycles. Thus, it was only at that age that O-M rats responded as rapidly as DA rats at the age of 70 days.

Hybrids between the O-M and DA strains presented intermediate results (Fig. 6). Ten F_1 hybrids were tested in continuous light for 10½ days (74–91 days old at the start). Fourteen continued to cycle, while 3 gave a PE response at only the very end of the period. Another group of 12 F_1 hybrids of similar age were in the light for 20½ days. None responded during the first 10 days, while 7 developed PE during the second 10 days. Similar results appeared among F_2 hybrids.

Hybridization thus shortened the response time in about half of the animals. It is significant that the DA strain, having by far the earliest age of onset of SPE, was also much more responsive to continuous light and that hybrids were intermediate between the parent strains with respect to both SPE and LLPE. To determine whether individuals that had been most responsive to continuous light would later show relatively early SPE, many of the hybrids were followed for several months. Although 21 of 49 previously illuminated rats eventually displayed SPE during the 200 to 260-day age range, correlation of the two phenomena in individual animals was not apparent. Nevertheless, there are distinct influences of both age and genetic background on the expression of both LLPE and SPE.

The occurrence of SPE in relatively young DA rats and their rapid LLPE response encouraged attempts to restore cyclic function. SPE ovaries were shown to luteinize in response to administration of anterior pituitary extract. Furthermore, the restoration of cycles by simply shortening the daily light ration was proof that appropriate regulatory mechanisms had not been lost. The search for these mechanisms led inevitably to: 1) disclosure that under some circumstances progesterone stimulates ovulation, facilitating the positive feedback action of estrogen; 2) evidence for a spontaneous acute surge of gonadotropin secretion, beginning during a predictable "critical period" on the afternoon of proestrus; 3) evidence that this surge can be blocked by various drugs having central neural action; 4) demonstration of 24-hour periodicity in the apparatus controlling the surge; 5) demonstration that stimulation of the medial preoptic area (MPOA) or septum can induce an ovulatory surge of gonadotropin, that the surge can be induced a) in proestrous rats blocked pharmacologically, b) in 5-day rats during late diestrus, and c) during SPE or LLPE; 6) the concept of dual "centers" in the brain controlling gonadotropin secretion in the female rat, with cyclic control of ovulation involving a diffuse septo-preoptic-tuberal neuronal system; 7) evidence for separate neural mechanisms inducing ovulation and pseudopregnancy in this species; 8) "delayed pseudopregnancy"; and 9) demonstration that separation of the pituitary gland from its normal relationship to the brain enhances prolactin secretion, other secretions of the pars distalis being curtailed, but capable of being restored through renewed vascular connection from the median eminence.

Steroid Regulation of the Ovarian Cycle

The inhibitory chronic effects of gonadal secretions on pituitary gonadotropic function were recognized several years before the synthetic steroids became available (cf. Burrows 1949). These negative effects were the basis for the "push-pull" hypothesis of pituitary-gonadal interaction proposed by Moore and Price (1930, 1932). A positive, stimulative effect of gonadal hormones in promoting puberty was noted by Engle (1931), however, and several workers soon reported that the immediate effect of estrogen administration on luteinizing potency of the AP was stimulative (Fevold, Hisaw and Greep 1936; Lane and Hisaw 1934; Lipschütz 1935). More directly, Hohlweg and Chamorro (1937) demonstrated that luteinization could be induced in immature rats by administration of estradiol benzoate. In ewes the induction of ovulation by estrogen therapy was accomplished by Hammond et al. (1942).

On the assumption that in SPE and LLPE, in spite of the apparent continuous secretion of estrogen the amount in circulation might be too low, I tried various treatments with estrogen in SPE DA rats, but without any indication of luteinizing action. It seemed more likely that progesterone would be of use, at least for interrupting the steady state in view of its well-known suppression of estrus. The effects of varied dosages of progesterone on the rat cycle had been reported by Phillips (1937). After demonstrating that daily subcutaneous injections of 1.5 mg of progesterone in oil suppressed estrous cycles, resembling the effect of corpora lutea during pseudopregnancy or pregnancy, he determined that smaller amounts allowed cycles to proceed: 1.0 mg/day tended only slightly to prolong diestrus, and even that effect was lost when dosage was further reduced.

Restoration and Maintenance of Ovulatory Cycles by Progesterone Treatment of Persistent-Estrous Rats

To test the ability of progesterone to counteract SPE, a preliminary exploration with different dose levels was carried out in DA rats, 9 young cyclic animals and 8 rats in which SPE had become established (Everett, unpublished). The first injection to the cycling rats was given on either the day of estrus or D-1. In agreement with Phillips, 1.5 mg progesterone per day consistently inhibited estrus; equivalent results were obtained in two rats with 1.25 mg doses. The effects in the SPE rats were somewhat different, for 3 rats receiving 1.25 or 1.5 mg daily presented one day of vaginal cornification after 4 or 5 days of leukocytic smears. In 2 rats corpora lutea of normal appearance, grossly and histologically, were present 2 or 3 days later. The possibility that even smaller amounts might lead more frequently to ovulatory cycles, proved to be true (Everett 1940).

The effects of low dosage are summarized in Figs. 7 and 8. In the great majority of trials, to interrupt SPE a standard dose of 1.0 mg progesterone was given for two days. Thereafter the dosage was lowered as illustrated. Whereas the interrupting doses were commonly given in the late afternoon, the other injections were made near the noon hour. This earlier treatment was fortunate, for later evidence shows significantly different effects of progesterone administration in early and late afternoon, respectively (Everett and Sawyer 1950; Redmond 1968; Zeilmaker 1966b): only the early injections tend to favor ovulation.

In the *short-term experiments* (Fig. 7), ovulation was nearly always induced by daily injection of 0.25 to 0.5 mg progesterone. Full sets of corpora lutea were present at terminal laparotomy in the ovaries of 7 of the 8 rats receiving treatment

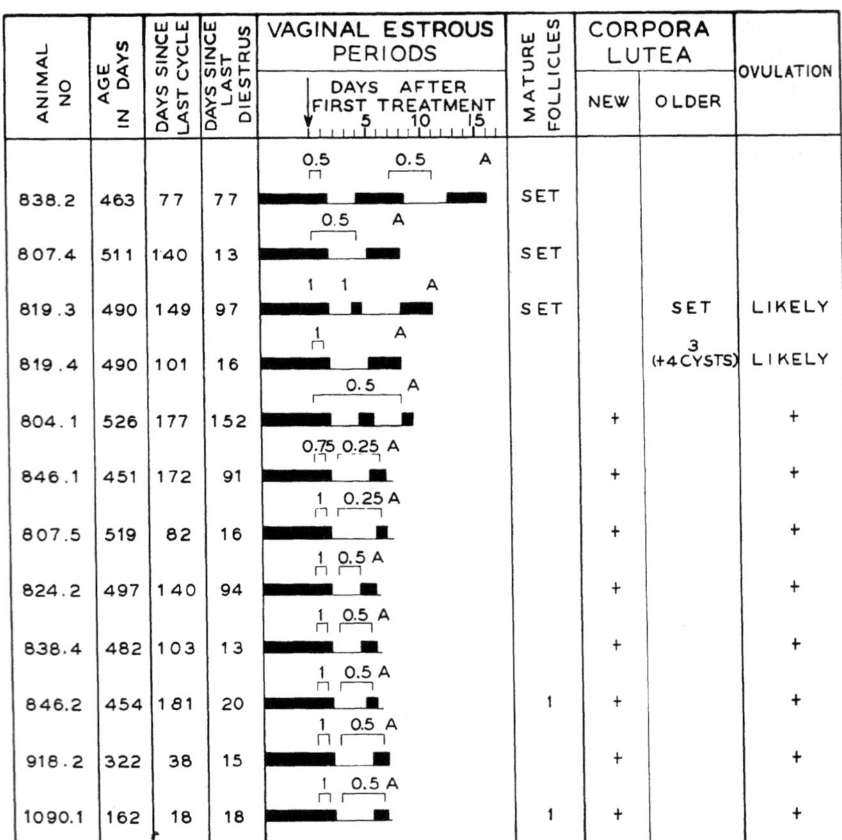

Fig. 7. Vaginal sequences in persistent-estrous rats treated for short periods with small amounts of progesterone. Black bars represent proestrus-estrus, base line diestrus. Progesterone dosages in mg/day are shown by the numerals above the brackets over each sequence. Ages etc. refer to the day of first treatment. A = autopsy. (From Everett 1940)

throughout diestrus and usually into the next estrus. The exceptional rat (#846.1) had only one ruptured follicle, two that luteinized without rupture, and seven atretic follicles. The first cycle of #804.1 was not accompanied by luteinization. When treatment ended before the final day of diestrus (#838.2 and #807.4), PE returned promptly and a few days later the ovaries contained new sets of Graafian follicles. In #819.3 a cyclic estrus was induced by the second injection of 1.0 mg progesterone in late diestrus. In #819.4 the presence of 3 well formed corpora lutea and 4 lutein cysts indicated that the interrupting injections themselves could sometimes induce luteinization.

In *long-term experiments* (Fig. 8), daily injection of progesterone continued for 3 to 7 weeks in dosages ranging from 0.1 to 1.0 mg. This usually brought about sequences of regular cycles, most of which were accompanied by corpus luteum formation, as judged at laparotomy or autopsy. One rat, not illustrated, while receiving only 0.15 mg progesterone per day, passed through 9 regular cycles and possessed at least 3 sets of corpora lutea during the final diestrus. Another rat (record

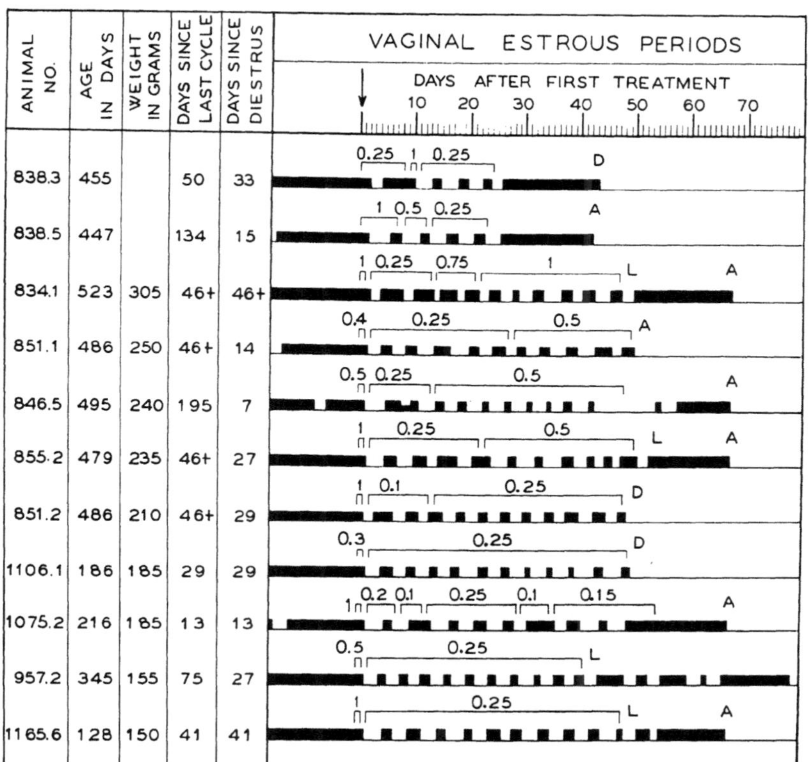

Fig. 8. Long-term vaginal sequences in persistent-estrous DA rats treated daily with low dosages of progesterone, showing consecutive cycles during the treatment and resumption of persistent estrus afterward. A = autopsy, D = discontinued or transferred to another experiment, L = laparotomy. Other details as in Fig. 7. (From Everett 1940)

not shown) failed to cycle while receiving only 0.1 mg progesterone for 22 days and had only two brief periods of diestrus of doubtful significance. In all rats, withdrawal of treatment resulted in prompt return of PE.

In *rats exposed to continuous light* (unpublished), LLPE was likewise prevented or interrupted by daily administration of small amounts of progesterone (Fig. 9). Fifteen DA females 94 to 104 days of age were divided into groups A and B, litter mates being apportioned as nearly equally as possible. All were in continuous light for 23 1/2 days. Each rat in group A received daily injection of 0.5 mg progesterone per 180 g body wt, beginning on day 11. The 8 rats in group B each received this treatment during the first 10 days, after which progesterone was withheld. During progesterone administration there was little prolongation of vaginal estrus in either group. In group B the PE response developed more rapidly after progesterone withdrawal than it had in group A before treatment began. Hence, in group B there appeared to be an underlying urge toward PE imparted by the constant illumination. As the ovaries were not examined in these experiments it is impossible to know whether luteinization was induced by the progesterone treatment, although it is most likely.

Individual records from some of the SPE experiments suggested that progesterone might be important primarily as estrus approaches. When treatment was omitted late in diestrus, the forthcoming estrus commonly persisted. On the other hand, daily treatment continuing to the end of diestrus produced new corpora lutea in two rats, and one rat luteinized after receiving only an isolated injection on the last day of diestrus. Experiments were carried out to determine the relative ovulatory effectiveness of single doses of progesterone injected on different days of diestrus, on the day of proestrus (vaginal stages I or II), or at progressively later times after a new period of SPE ensues (Tables 4 and 5) (Everett 1943). In early all cases SPE was first

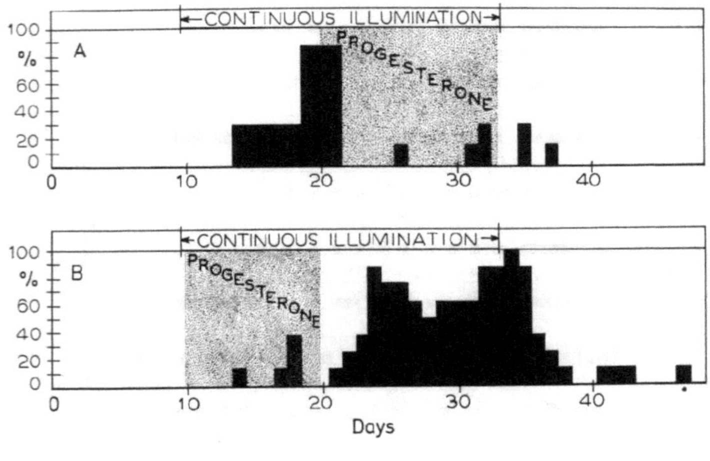

Fig. 9. Suppression of light-induced prolonged estrus (LLPE) during daily treatment with small amounts of progesterone (0.5 mg in oil sc). In black are the daily percentages of rats showing vaginal estrus longer than that of the normal cycle. DA rats

interrupted by an initial injection of 0.5 or 1.0 mg progesterone. In a few cases, advantage was taken of a spontaneous interruption. A single animal could often be used more than once. For example, if she received treatment on D-2 and nevertheless returned to PE, another injection could be given after renewed vaginal cornification had continued several days. The experiment was so designed that in each individual no more than three sets of corpora lutea would be formed, which could be distinguished histologically. The occurrence of ovulation was directly established by the presence of tubal ova in 18 of 20 cases in which luteinization had occurred within the past 48 hours. Otherwise, ovulation was indicated by the presence of corpora lutea having normal histologic appearance without sign of entrapped ova.

Control data demonstrated that very few SPE rats would spontaneously experience an ovulatory cycle after receiving only an interrupting dose of progesterone (incidence: 2 cases among 45 rats = 4.4%). The likelihood of a spontaneous interruption of SPE coincident with experimental treatment was estimated as between 2.5 and 5%.

Following interruption of SPE and injection of progesterone on D-I very few rats completed a cycle and luteinized. However, injection of the second dose of progesterone on day 2 of a 2-day interval produced cycles and corpora lutea in 5 of 15 cases (Table 4). Injection of the second dose on D-2 or D-3 of a 3-day interval

Table 4. Relative effectiveness of single doses of progesterone given on different days of diestrus or on the first day of renewed estrus after interruption of SPE. DA rats. (From Everett 1943)

Day of administration of test dose	Dosage level, MG.	No. of cases	Interval followed by cycle No.	%	Corpus Luteum formation No. cases	%
A. 2-day interval						
Diestrus, *day 1*	0.5–1.0	1	1		1	
Diestrus, *day*	0.5–1.0	15	5	33	5	33
Estrus, *day 1 (stages I,II)*	0.5–1.0	22	22	100	13	59
	2.0	8	8	100	2	25
Totals		46	36	78	21	46
B. 3-day interval						
Diestrus, *day 1*	0.5–1.0	10	2	20	2	20
Diestrus, *day 2*	0.5–1.0	13	9	64	9	64
Diestrus, *day 3*	0.5–1.0	15	11	73	10	67
Estrus, *day 1 (stages I,II)*	0.5–1.0	24	24	100	19	79
	2.0	17	17	100	12	71
Totals		79	63	80	52	66

Table 5. Declining effectiveness of single doses of progesterone (0.5 to 1.0 mg) as renewed estrus persists. (From Everett 1943)

Day of estrus	Number of cases	Corpus luteum formation	
		Number of cases	%
1[a]	46	32	70
2–3 inclusive	19	7	37
4–5 inclusive	19	5	26
6–7 inclusive	19	3	16
8–9 inclusive	17	2	12
10–100 inclusive	79	8	10

[a] Combined data from Table 4, A and B.

completed the cycle and caused luteinization in 70% of the cases (19/28). Injection during proestrus always completed the cycle and caused corpus luteum formation in 70% of the rats (32/46). When estrus was allowed to continue for awhile, the effectiveness of the second treatment with progesterone declined progressively: after the first week the frequency of luteinization remained about 10% (Table 5).

The conclusion was obvious: progesterone administration strongly favors ovulation and luteinization when given during late diestrus and early estrus. Interaction with estrogen was indicated.

Advancement of Ovulation by Progesterone in Normally Cycling Rats

Boling et al. (1941) in their study of follicle growth in the rat ovary during the estrous cycle reported that at the beginning of preovulatory swelling the follicle volume is a linear function of the length of the cycle. In 4-day cycles ovulation occurs from smaller follicles than in 5-day cycles. The authors interpreted this to mean that in the 5-day cycle the follicles are ready for the luteinizing stimulus a full day before it arrives. The above evidence from SPE rats treated with progesterone suggested a means by which such a stimulus might be provoked prematurely.

Two groups of 5-day cyclic rats were examined (Everett 1944). The control group of 14 animals comprised 5 O-M rats, 2 DA rats and 7 hybrids. Each had history of several 5-day cycles in succession and presented a proestrous vaginal smear at the expected time in the current cycle. On this day, the ovaries contained large Graafian follicles without evidence of prelutein change. Approximately half of the controls received 0.4 ml sesame oil on D-3. The experimental group of 19 rats comprised 6 O-M rats, 5 DA rats and 8 hybrids, selected like the controls. Each received either 1.0 or 2.0 mg progesterone about noon on D-3. On the next day, the vaginal smear had advanced to stage II in all but 4 rats. In 16 of the 19 rats, the ovaries on that day contained freshly forming corpora lutea. In most cases there were ova in the ampullae of the oviducts. Thus, indeed, progesterone had advanced ovulation time about 24 hours as predicted. In 4-day cyclic rats similarly treated on D-2, there was no advance (Fig. 10B).

However, it was possible to transform the 4-day cycle into an artificial 5-day cycle by giving 1.0 or 1.5 mg progesterone on D-1. This effect was obtained uniformly in 18 O-M rats receiving the 1.5 mg dose (Fig. 10D); ovulation was delayed 24 hours. Injection of progesterone on both D-1 and D-2 brought about a delay of 48 hours (Fig. 10C).

Just as in the normal 5-day cycle, progesterone injected on D-3 of the artificial 5-day cycle advanced ovulation time (Fig. 10E). Five O-M rats with histories of 4-day cycles were injected with 1.5 mg progesterone on D-1. Two days later, when the vaginal smears were still diestrous, each rat received 1.0 mg progesterone between 1200 h and 1300 h. On the following day when four were killed, all had fresh corpora lutea and tubal ova. The fifth rat, killed on the next day had a solitary corpus luteum, judged histologically to be about 30 hours old, together with many atretic follicles. *Here was clear evidence of a biphasic action of progesterone, initially to stimulate and later to retard.*

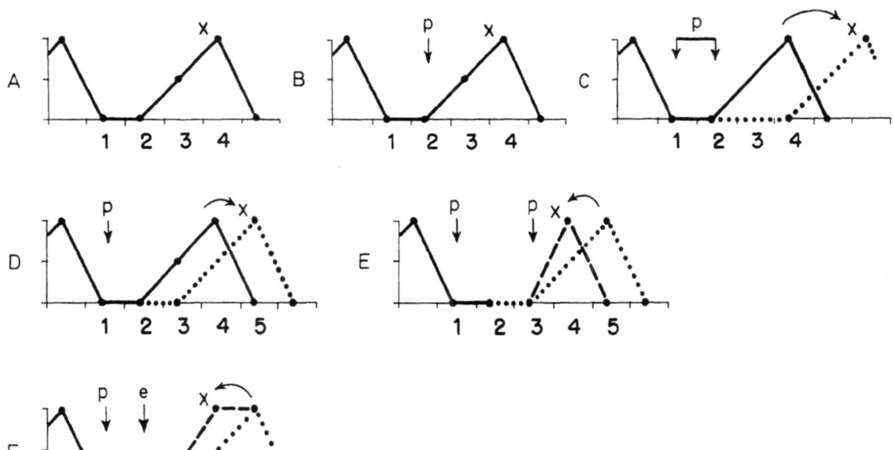

Fig. 10. The 4-day estrous cycle and experimental modifications. Two units of the ordinate represent a full estrous smear, each unit on the abscissa represents 24 h, midnight to midnight. Days of the cycle are numbered progressively from the first day of diestrus. Approximate ovulation time is marked by "X". O-M rats. **A** The standard cycle: ovulation early on day 4. **B** Progesterone (p) injected on diestrus day 2 does not advance ovulation and may block it. **C** 1.5 mg Progesterone on diestrus days 1 and 2 retards the cycle 2 days. **D** An artificial 5-day cycle produced by giving 1.5 mg progesterone on diestrus day 1: vaginal estrus and ovulation retarded one day. **E** In such an artificial 5-day cycle, as in the normal 5-day cycle, additional progesterone on diestrus-3 advances estrus and ovulation. **F** In the artificial cycle, estradiol benzoate (e) injected on diestrus day 2 advances ovulation as it does in the normal 5-day cycle. (From Everett 1948)

The biphasic action was later documented extensively in the rabbit (Kawakami and Sawyer 1959; Sawyer and Everett 1959), with respect to certain relevant thresholds to electrical stimulation of the central nervous system. The rabbit studies disclosed that for a few hours after administration of progesterone to an estrous doe, the effect is to facilitate. That is succeeded by a depressant phase, which continues on the following day.

Thus, in rats, when progesterone is administered daily in amounts exceeding 1.0 mg, the depressant effect of the first dose is reinforced and extended from day to day. Without reinforcement, that effect is lost by D-3; progesterone can then exert its stimulative action, subject to renewed priming by estrogen. Apparently when the daily supply of progesterone is low the reduction of gonadotropin secretion is inadequate to prevent the rise of new follicles and renewed estrogen priming. Here again we see the facilitative power of progesterone. Further discussion awaits presentation of experiments showing that estrogen can induce ovulation during pseudopregnancy or pregnancy (p. 33), in the face of high levels of progesterone.

Corpora Lutea of Cyclic Rats as Possible Sources of Progesterone

Comparison of the ovaries of cyclic DA rats with those of the O-M strain disclosed a pronounced difference in the histological and histochemical characteristics in the corpora lutea from the preceding cycle when examined in late diestrus, proestrus and on the day of estrus after the new ovulation. The differences in histology are summarized in Figs. 11 and 12 from Everett (1945). In the (normal) O-M corpora lutea one sees evidence at all these stages of greater activity than in the corresponding stages of DA corpora lutea. The individual luteal cells were larger, lipid inclusions more numerous, and islands of fatty necrosis at proestrus and estrus far more extensive. At proestrus the cells from areas outside these islands resembled those in fully active corpora lutea of pseudopregnancy (Fig. 11). At estrus there was a general increase of lipid, by sharp contrast with the typical DA corpora lutea at estrus. Application of the Schultz histochemical test to frozen sections of formalin-fixed corpora lutea presented the characteristic blue-green color of cholesterol and its

→

Fig. 11. Sudan III stains of normal corpora lutea of the estrous cycle (O-M strain). Frozen sections of formalin-fixed material, stained concurrently in the same dye solution. Sudanophil (lipid) elements in black and shades of gray. Prints matched in tone with original specimens. Representative cells of each section are shown in drawings at the right. 60X and 830X. (After Everett 1945). *Diestrus day 2*, presumptive 5-day cycle. Cell A is average. *Diestrus day 3*, 5-day cycle. Most cells range from type A to type C. Type B is rare. *Proestrus*, 5-day cycle. Marked lipid accumulation in irregular patches. Elsewhere the parenchyma has lost much of its lipid. Large sudanophobes like A are abundant, also intermediates between A and B. *Estrus*, 4-day cycle, post ovulatory. Lipid is abundant throughout. Interior of necrotic patch has many degenerate cells and bodies like C. Cells like D and E abound near the patch, while types A and B and intermediates characterize the outlying healthier parenchyma

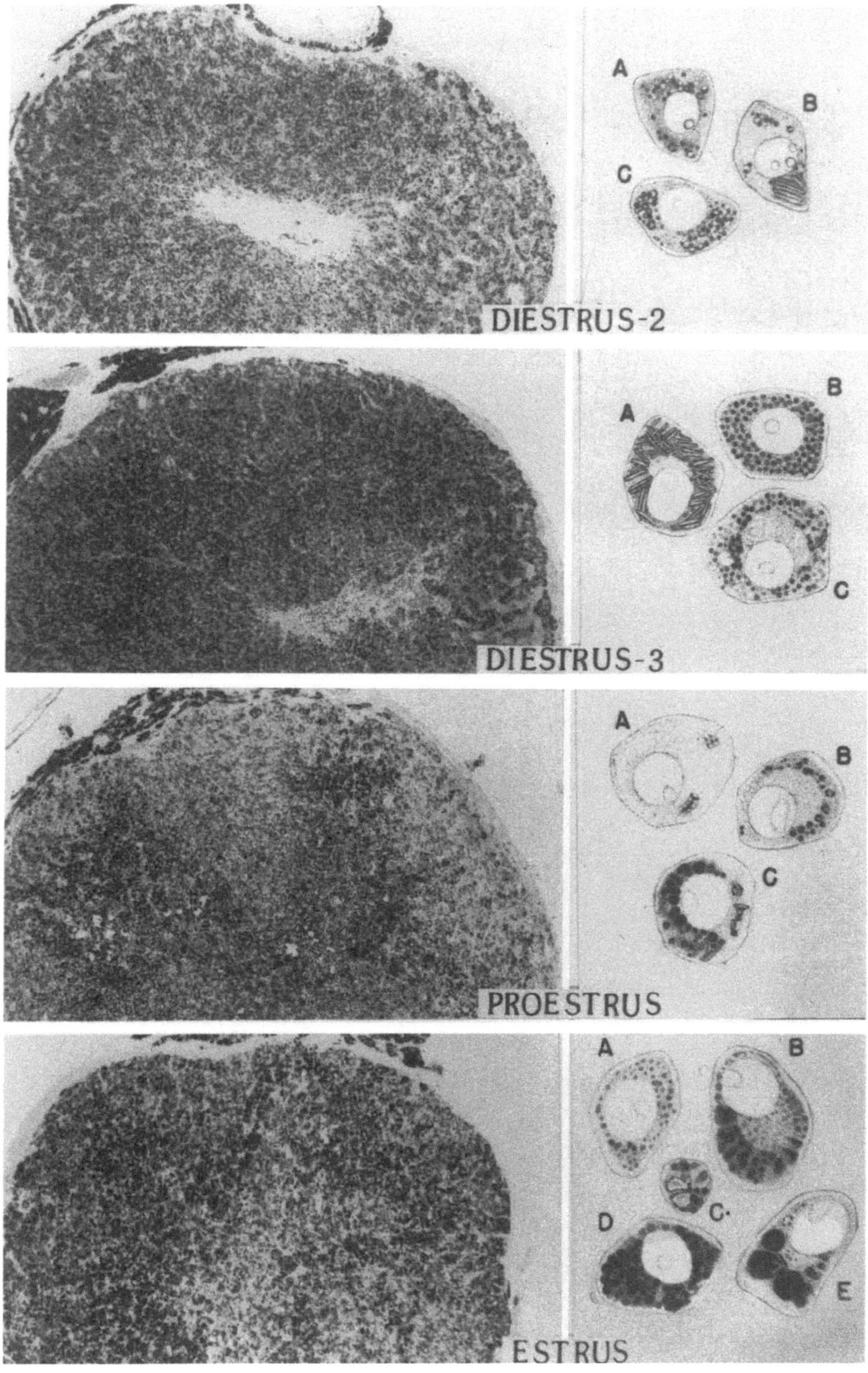
A
B
C
DIESTRUS-2
A
B
C
DIESTRUS-3
A
B
C
PROESTRUS
A
B
C
D
E
ESTRUS

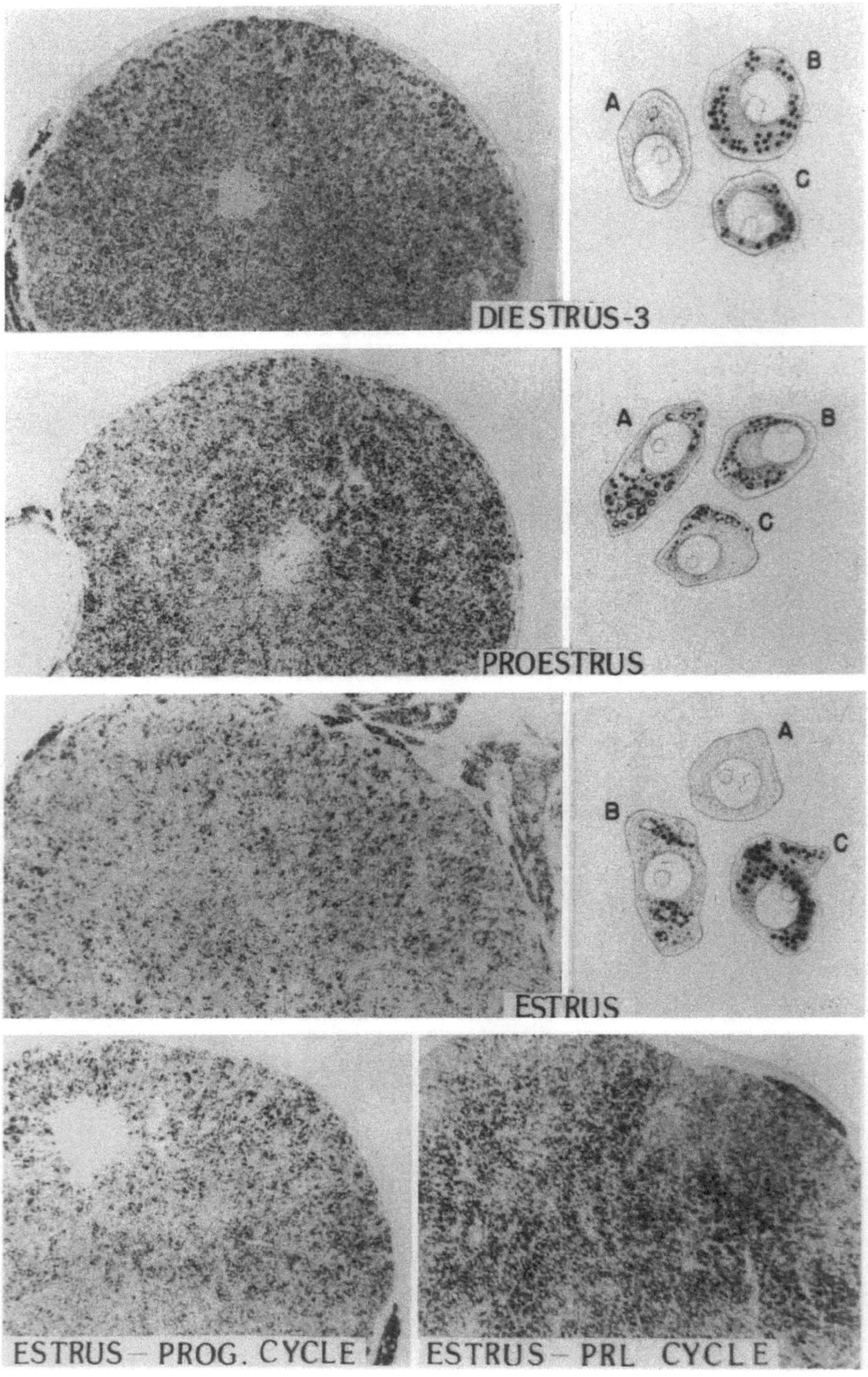
A
B
C
DIESTRUS-3
A
B
C
PROESTRUS
A
B
C
ESTRUS
ESTRUS - PROG. CYCLE
ESTRUS - PRL CYCLE

esters in the lipid-rich cells of O-M corpora lutea. DA corpora lutea were relatively lacking in cholesterol, the cells were smaller and the islands of fatty necrosis were reduced or absent. However, it proved possible by treatment with low levels of PRL to transform DA corpora lutea to make them resemble those from O-M rats.

The rationale and execution of these prolactin studies (Everett 1944b) were as follows. The fact that PRL is the principal luteotropic hormone in the pseudopregnant and early pregnant rat was evident from reports by Astwood (1941), Cutuly (1941) and Evans et al. (1941). The relative enlargement of cells in O-M cyclic corpora lutea suggested exposure to some moderate degree of PRL stimulation during the cycle, while DA rats lacked that stimulating action. It seemed reasonable that treatment of DA rats with small amounts of PRL below the amounts needed for full luteal secretion, would modify their corpora lutea to resemble those in O-M rats. Furthermore, it seemed possible that if corpora lutea were first introduced in DA SPE rats by a cycle of progesterone treatment they could be stimulated by low levels of PRL to produce enough progesterone intrinsically to bring about ovulation in the next cycle. From data of Evans et al. (1941) and Tobin (1942) it was estimated that to sustain full luteal function the minimum daily dosage of PRL should be 15–20 IU and that the resulting level of progesterone secreted would approximate that supplied by 1.5 mg progesterone injected once daily. Hence, to reproduce the effects of low-level progesterone the PRL dosage should be reduced proportionately.

The experiments were carried out with PRL supplied by the Schering Corporation. Significant contamination with LH in this material was ruled out by the fact that, in the absence of luteal tissue, PRL dosages as high as 30 IU failed to induce ovulation or even to produce histological evidence of luteinization. Furthermore, in hypophysectomized rats there was no repair of the interstitial tissue. It was apparent that PRL could not stimulate significant progesterone secretion from Graafian follicles or other non-luteal ovarian tissue. On the other hand, when corpora lutea were first introduced by progesterone treatment, certain PRL regimens produced sequences of ovarian cycles, as illustrated in Fig. 13. Among 56 attempts to induce the initial set of corpora lutea by progesterone treatment, 40 (71%) were successful. In the principal experimental series (Fig. 13e, f, and g), among 22 potential cycles during PRL treatment, 16 (73%) were completed, each with formation of new corpora lutea. Whenever luteinization failed the animal returned to PE. The rate of failure was similar to that expected from progesterone treatment itself (see p. 24). The obvious conclusion is that PRL had induced moderate secretion of progesterone

◀

Fig. 12. DA corpora lutea, stained with Sudan III concurrently with those in the preceding figure. 49X and 682X. (After Everett 1945). *Diestrus day 3*, 5-day cycle. Most cells are like A and C. Type B is rare. *Proestrus*, 5-day cycle. Most cells are between types B and C. *Estrus*, 5-day cycle. Sudanophilia is slight, most cells ranging between types A and B, tending toward A. Type C is rare. *Full Estrus* after a progesterone-induced cycle in a SPE rat essentially identical to the above. Extensive patches of fatty necrosis, elsewhere most cells are like cell A in Fig. 11 Estrus and are distinctly larger than in the 'progesterone cycle'. *Full Estrus* after a 'prolactin cycle' (see Fig. 14)

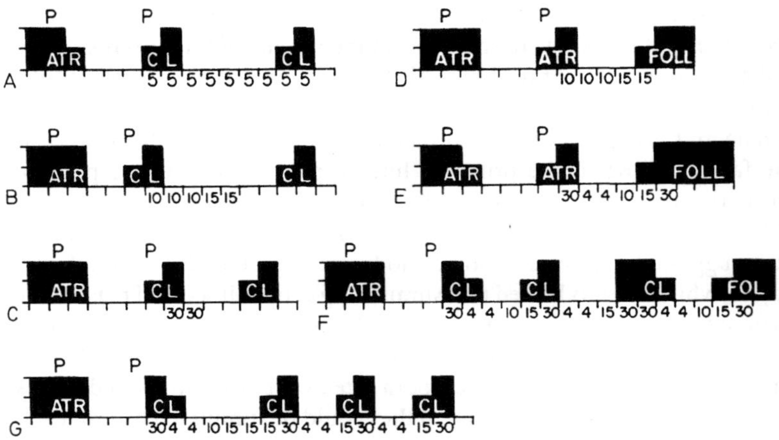

Fig. 13. Typical experiments showing maintenance of vaginal cycles in spontaneously persistent-estrous DA rats by small daily doses of prolactin after initial induction of corpora lutea by progesterone treatment. Black areas represent the degrees of vaginal estrus: one division of the ordinate = proestrus or metestrus; two divisions = full estrus. Each unit of the abscissa indicates one day. P = injection of progesterone. Prolactin dosage in international units is shown by numerals below the base lines. ATR = atresia of large follicles; CL = corpus luteum formation. (From Everett 1944)

from the current set of corpora lutea and that progesterone in turn had brought about the ovulatory discharge of gonadotropin.

Histological and histochemical examination of ovaries removed on the day after ovulation in a "PRL cycle" disclosed corpora lutea closely resembling those in normally cycling O-M rats (Fig. 14). Individual cells were enlarged and most now had foamy cytoplasm indicative of lipid inclusions. Frozen sections of representative formalin-fixed corpora lutea were strongly positive for cholesterol in the Schultz test.

The several findings outlined above, both physiological and histological, implied that in the normal rat the corpora lutea may become slightly active during the estrous cycle and may constitute a source of progesterone promoting the ovulatory discharge of LH. Two papers (Everett 1980, 1984) examined this possibility further, showing that in middle-aged SPE rats of the CD strain the induction of a set of corpora lutea by an injection of LH is usually followed by several ovulatory short cycles (Fig. 15c and d) after which persistent estrus resumes. Aschheim (1965) had shown that such treatment of *senile* SPE rats resulted immediately in pseudopregnancy which was then followed typically by a number of short cycles and eventual anestrus. He interpreted this to mean that the pseudopregnancy was an essential precondition for the short cycles. Treatment with ergocornine, a known inhibitor of prolactin secretion, prevented the pseudopregnancy and the animal returned promptly to PE. A few experiments with middle-aged CD SPE rats by Zeilmaker and Everett (1967, unpublished) showed that a number of short cycles followed immediately after an LH injection, but the full significance was not realized at that time.

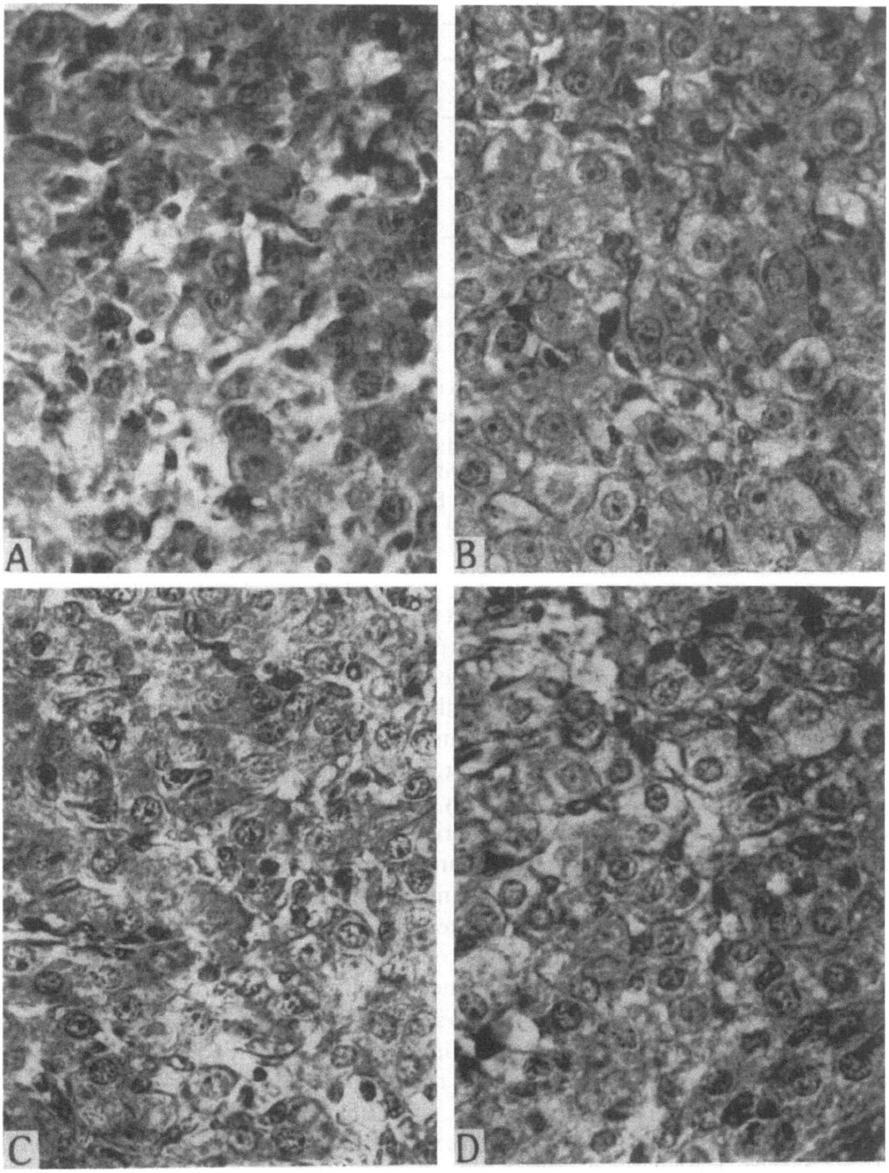

Fig. 14. Typical areas of corpora lutea from different types of rats. **A** Cyclic DA rat. Corpus luteum of the next youngest set. Diestrus day 1. **B** Normal O-M rat. Corpus luteum of similar age to that in A. The cells are larger and vacuolation makes cell boundaries more distinct. **C** Persistent-estrous DA rat rendered cyclic by progesterone. Corpus luteum removed on diestrus day 1 of the second cycle. **D** Persistent-estrous DA rat after 3 cycles under prolactin treatment (Fig. 13G). Diestrus day 1 after final cycle. Increased cytoplasmic vacuolation by contrast with C, resembling the normal corpus luteum (B). (From Everett 1944)

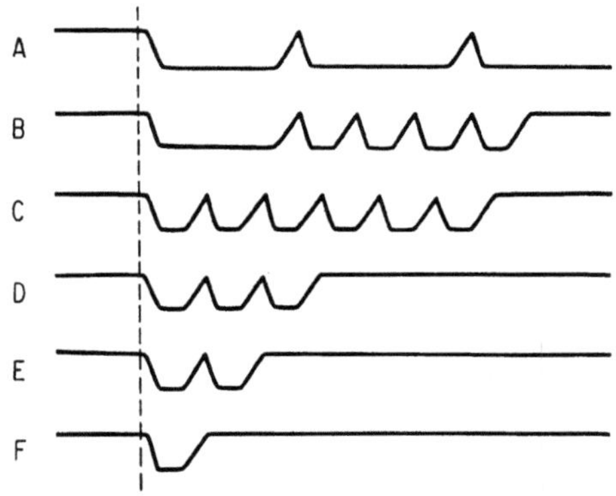

Fig. 15. Varied results of injection of an ovulatory dose of LH to spontaneously persistent-estrous CD rats. Injection time is marked by the broken line. Vaginal cornification in each case is indicated by the upper level and diestrus by the lower level. Vaginas consistently diestrous 2 days after LH injection. A = pseudopregnancies in repetition; B = several short cycles after an initial pseudopregnancy; C to E = one or more short cycles before return to SPE. F = rare, immediate return to SPE. See Tables 6 and 7. (From Everett 1980)

I was able later (Everett 1980) to show that this was the common response (Fig. 15) of middle-aged SPE subjects, whereas initial pseudopregnancies were more usual in rats over a year old (Table 6 and 7). Aside from the very few rats that presented pseudopregnancies in succession, all rats that were kept long enough returned to the PE state instead of gradually passing into anestrus as had Aschheim's senile rats. Treatment of SPE CD rats with either of two ergot alkaloids during the diestrus that immediately followed LH injection nearly always resulted in failure to complete the cycle, PE returning after the short interval (Everett 1980, 1984). Bromocryptine (CB-154, 2-bromo-α-ergocryptine methane sulfonate) in a fine aqueous suspension 0.5 mg/0.5 ml/100 g BW, injected subcutaneously mid-afternoon of the first or second day of diestrus, nearly always prevented the cycle (Fig. 16). The same result occurred after treatment with lergotryl mesylate (LRG, 2-chloro-6-methylergoline-8β-acetonitrile methane sulphate) in similar dosage of aqueous solution injected intraperitoncally daily for 3 days beginning on D-1. The blocking action of bromocryptine was overcome by administration of ovine PRL at the same time as the ergotamine on D-2 and sometimes repeated on D-3 (Tables 8 and 9). PRL on D-3 was evidently superfluous, however, and usually prolonged the diestrus to 5 days instead of 2–3 days. Table 10 shows the serum prolactin concentrations at the time of the LH injections in relation to the type of response, whether an initial pseudopregnancy, one or more cycles, or direct return to PE. Although the data strongly suggest a positive relationship between the initial PRL level and the degree of corpus luteum activation in the days following, large variances denied statistical significance.

A considerable body of experimental evidence from other laboratories now leaves no doubt that rat corpora lutea are moderately active in progesterone secretion during the estrous cycle (de Greef and van der Schoot 1979; Rothchild 1981). Assay

Table 6. Types of response to injection of an ovulating dose of LH in spontaneously persistent estrous CD rats according to age at time of treatment. (From Everett 1980)

Response pattern	Respective numbers of rats	
	6–11 months old	12–17 months old
PSP → PSP → PSP → PSP	1	3
PSP[a] → cycle(s) → (SPE)[b]	9	4
Cycle(s) → (SPE)[b]	30	2
SPE	6	1
Totals	46	10

[a] PSP was proven in five of five additional rats by decidual reactions to threads placed in the left uteri on day 4 of leukocytic vaginal smears.
[b] Return to SPE when followed.

Table 7. Comparative numbers of short cycles intervening before the reappearance of persistent estrus in two groups of rats. (From Everett 1980)

Pattern of response to LH injection	n	No. of Cycles				
		1	2	3	4	>4
PSP → cycle(s) → SPE	12	1	0	1	1	9
Cycle(s) → SPE	26	9	5	1	4	7

See Figs. 15 and 16, Tables 6, 8–10

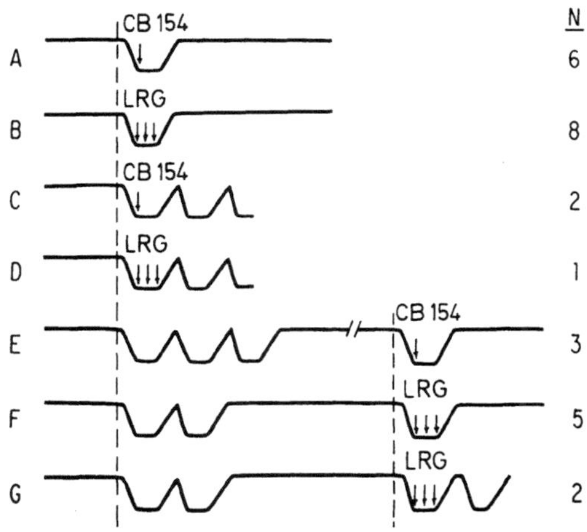

Fig. 16. Treatment with ergotamines during the initial diestrus after LH-induced luteinization usually prevented cycling. N = number of rats in each category. LH injection marked by broken line. Arrows mark injection of CB-154 or Lergotril. (From Everett 1980)

Table 8. Effectiveness of prolactin in overcoming blockage by bromocriptine of estrous cycles in persistent-estrous rats luteinized with LH. (From Everett 1984). (See Fig. 16)

Bromocriptine injection (0.5 mg)[e]	Prolactin treatment[e]		No. of rats initially luteinized	Results	
	D-2[c]	D-3		Cycle completed	Return of SPE[d]
D-2	5.0 i.u	5.0 i.u.	3	3	
D-2	2.5 i.u.	2.5 i.u.	11	8	3
D-2	2.5 i.u.		6	4	2
D-2			6	1	5
D-1			11[a]	2	9
			70[ab]	59	11

[a] Includes data from Everett (1980).
[b] Excludes 21 rats that immediately became pseudopregnant.
[c] D-1, D-2, D-3: diestrus days 1, 2 and 3, respectively.
[d] SPE: spontaneous persistent estrus.
[e] Doses of bromocriptine and prolactin are per 100 g body wt, sc.

Table 9. Duration of diestrus in the several treatment groups of rats. (From Everett 1984)

Treatment	Nos. of rats distributed by length of diestrus in days: 2	3	4	5	6	7	Total nos. of rats	Mean length of diestrus (days ± SEm)
LH only	13	33	13	6	2	1	68	3.32 ± 0.13
LH Bromocriptine	5	11	1				17	2.76 ± 0.14
LH Bromocriptine PRL, 1 day		3	3				6	3.50 ± 0.04
LH Bromocriptine PRL, 2 days		2	1	11			14	4.64 ± 0.21

PRL: prolactin.

Table 10. Relationship between type of response of persistent estrous rats to LH administration and serum prolactin concentration at time of injection. (From Everett 1984)

Type of response	No. of rats	Prolactin concentration (ng/ml serum)[a]
Pseudopregnancy	6	235 ± 77
Multiple cycles	8	186 ± 30
Single cycle	6	88 ± 27
Persistent estrus	2	99,235
	22	171 ± 24

[a] In terms of NIADDK Rat Prolactin RP-1.

of circulating progesterone in both 4-day and 5-day cycles discloses not only a marked elevation on proestrus coincident with the LH surge, but significant increases beginning D-1 ("metestrus") and extending into the next day, reaching baseline on the D-2 afternoon in 4-day cycles or on the D-3 morning in 5-day cycles (Nequin et al. 1979; Roser and Bloch 1971). During the first 24–36 hours after ovulation the secretion appears to be autonomous, independent of luteotropic influence by either PRL or LH (Rothchild 1981; Sanchez-Criado et al. 1986; van der Schoot and Uilenbroek 1983). Bromocryptine injection at estrus in 5-day cyclic rats, eliminating PRL, shortens diestrus by one day, accompanied by curtailed progesterone secretion (Heuson et al. 1970; van der Schoot and Uilenbroek 1983). By contrast, 4-day rats are unaffected by that treatment (Aron 1979; Boehm et al. 1984). Rats subjected to daily treatment with bromocryptine, hence devoid of significant PRL secretion, presented series of 4-day cycles (Dohler and Wuttke 1974). It is important to recognize that the studies cited above have been carried out in young adult rats. Caution is necessary in applying the data to older animals.

The restoration of cyclic ovulatory surges of LH by progesterone in SPE rats has long been interpreted as showing a protective action of the steroid through enhancing the stimulative effect of estrogen (Everett 1961). The return of spontaneous cycling after introduction of just one new set of corpora lutea expresses the same protective effect. That this is the result of moderately elevated PRL is highly likely, for the protection is lost when the rats receive ergotamines during the initial diestrus. Whatever progesterone may be secreted autonomously on the first day or so is not usually effective in giving protection; it must continue to be produced into the final day of diestrus (or perhaps into the proestrus) to ensure an effective LH surge. That may not always be true, for there were a few instances of short cycles occurring in spite of the ergotamine treatment (5/27 = 18.5%, Everett 1980). In the majority of cases, however, one must assume that there is a continuing low level of PRL production as if from incipient microtumors of the AP.

On that assumption, one is faced with a number of questions. Why does the protective effect wane, so that after one or a few cycles SPE returns? A corollary question arises in those SPE rats that initially show pseudopregnancy after LH injection: why does this pseudopregnancy usually give way to a number of short cycles and these, in turn, to SPE? Does PRL output gradually diminish? If so, is there perhaps regression of PRL-secreting microtumors? Or, does the return of SPE mark random reduction of the competence of a particular set of corpora lutea, resulting in inadequate progesterone production at a critical time? On the other hand, we have seen (p. 22) that for reasons unknown progesterone itself occasionally failed to invoke ovulation (Tables 4 and 5).

Induction of Ovulation by Estrogen

The investigation of corpus luteum cholesterol in cyclic rats led to studies with pregnant and pseudopregnant rats in the course of which estradiol benzoate was injected on the fourth successive day of vaginal diestrus (L-4) or later (Everett 1947). This treatment resulted two days later not only in cholesterol accumulation in the corpora lutea, but in ovulation as well. Both effects could be prevented by hypo-

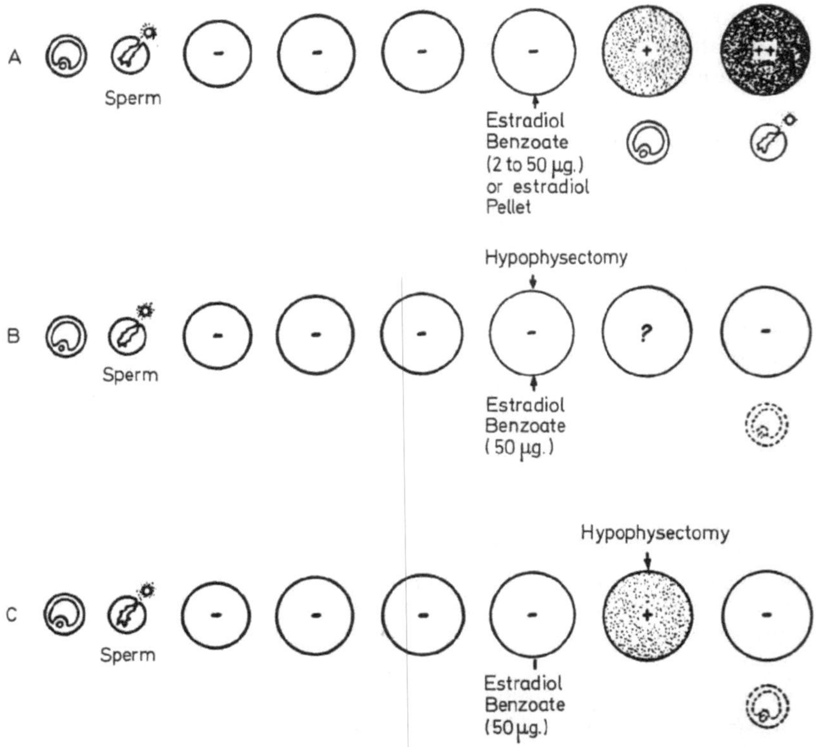

Fig. 17. Representation of experiments showing cholesterol accumulation in pregnancy corpora lutea accompanying ovulation after estrogen administration (A) and absence of these effects in hypophysectomized rats (B and C). Each symbol represents one day beginning with proestrus at left. Dotted outline of follicles in B and C indicate atresia. The relative amounts of cholesterol shown by the Schultz test on frozen sections are indicated by –, + and + +. (From Everett 1947)

physectomy on either the day of estrogen injection or the day following (Fig. 17). In the initial experiments, a standard subcutaneous dose of 50 μg estradiol benzoate (EB) in oil was used, but considerably smaller amounts (3 μg or more) proved to be equally productive of both effects (Table 11). The pseudopregnancy data were later enlarged, with 10 μg or 50 μg EB injected on L-4 after electrical stimulation of the cervix on the morning of estrus (Everett and Nichols 1968). In 6 of 6 rats this resulted in ovulation of 7 to 12 ova counted in the oviducts two days after the estrogen injection.

These observations have been repeated in other breeds of rats (Caligaris et al. 1972) with certain differences. Day 4 may not always be the most appropriate time for EB administration. Brown-Grant (1969a) reported that injection on L-4 usually failed to induce ovulation, while injection on L-5 gave good results. The difference was related to the presence or absence of competent follicles. Similar results have

Table 11. Similarity of dose-response relationships for ovulation and corpus luteum cholesterol accumulation in pregnant rats treated with estradiol benzoate. (From Everett 1947). (See Fig. 17)

No. of rats	Amount estradiol benzoate injected	Ovulation	Schultz test of corpora lutea
1	25 μg.	+	+ + to + + +
2	10 μg.	+	+ + to + + +
2	3 μg.	+	+ + to + + +
1	2 μg.	1 ovulated foll.	+ + to + + +
1	2 μg.	–	+ + to + + +
1	1 μg.	–	+
1	1 μg.	–	–
2	0.2 μg.	–	–

occurred in this laboratory (Everett and Tyrey 1982a). In either pregnancy or pseudopregnancy, CD rats receiving EB on L-4 usually failed to ovulate; yet their corpora lutea became distinctly fatty, indicative of the cholesterol response to an LH surge as noted above. Injection of 15 μg EB on L-5 or L-6 of pseudopregnancy produced ovulation in 2/2 and 4/5 rats, respectively. Inbred Fischer 344 rats (Charles River CDF strain) were singularly unresponsive to 15 or 50 μg EB injected on L-4, L-5, or L-6 of pseudopregnancy, ovulation failing uniformly in 6, 4, and 4 rats, respectively (unpublished). The presence of fatty corpora lutea in these negative cases suggests again that LH had been released, but that competent follicles were lacking.

Ovulation Advanced by Estrogen in Cyclic Rats

The fact that estrogen can induce ovulation in pregnant or pseudopregnant rats led to attempts to advance ovulation time in 5-day cyclic rats by giving EB (Everett 1948). Preliminary trials in 4 O-M rats disclosed that 33 to 100 μg EB injected on D-3 failed to advance ovulation, which occurred as normally expected during the second night after injection (Fig. 18). However, early ovulation was achieved consistently by estrogen administration on D-2, either 50 μg EB sc (5 cases) or small crystals of estradiol-17β (2 cases) implanted subcutaneously between 1100h and 1500h. Artificial 5-day cycles were produced in 5 other O-M rats. Two were injected with 50 μg EB and three with 15 μg EB on D-2. All 12 rats presented full sets of tubal ova 48 hours later and had appropriate numbers of newly forming corpora lutea. The vaginal smears were fully cornified. Similar results have been reported by Brown-Grant (1967, 1969) and Weick and Davidson (1970).

Ovulation can even be advanced 24 hours in 4-day cycles by administering EB on D-1 (Krey and Everett 1973). CD females received 5, 10, or 20 μg EB at various hours

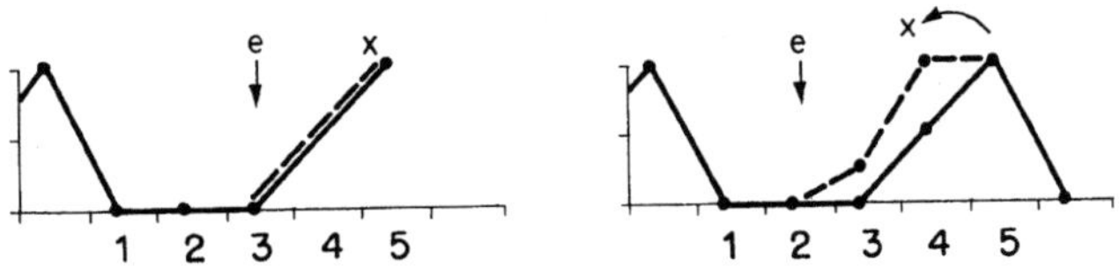

Fig. 18. Advanced ovulation in 5-day cyclic O-M rats induced by injection of estradiol benzoate (e) or implantation of small crystals of estradiol-17β on diestrus day 2. Two units of the ordinate indicate full estrus. Days of the cycle numbered consecutively beginning with diestrus day 1. Ovulation time indicated by "x". (From Everett 1948)

Table 12. Effectiveness of estradiol benzoate (EB) on day 1 in advancing ovulation 24 hr in 4-day cycling CD rats. (From Krey & Everett 1973)

EB treatment day 1		Results: Ovulation frequency[a] day 3		Luteinization frequency[b] day 3
5 μg	0500	1/6	(6.0 ova)[c]	1/6
	0700	2/6	(6.0 ova)	3/6
	0900–1000	2/6	(2.0 ova)	4/6
10 μg	0500	1/10	(3.0 ova)	5/10
	0700	5/10	(5.0 ova)	6/10
	0900–1000	6/10	(8.7 ova)	7/10
	1230	1/10	(2.0 ova)	3/10
	1500–1600	1/10	(7.0 ova)	1/10
20 μg	0500	3/10	(3.7 ova)	5/10
	0700	4/10	(7.8 ova)	7/10
	0900–1000	6/10	(4.2 ova)	8/10
	1230	8/10	(5.5 ova)	10/10
	1500–1600	6/10	(7.7 ova)	6/10
	1900	2/10	(13.0 ova)	2/10

[a] Ovulation frequency: rats ovulating day 3/total rats treated.
[b] Luteinization frequency: rats ovulating + rats with partial luteinization day 3/total rats treated.
[c] Mean ovum count/ovulating rat.

on D-1 from 0500h to 1900h (Table 12). The most effective regimen was a dose of 20 μg at about 1230h, resulting in the presence of tubal ova two days later in 8/10 rats and partial luteinization in the other two. The frequencies of advanced ovulation were almost as good as this after injection of 20 μg at either 0900h-1000h or 1500h-1600h. Kobayashi et al. (1971) and Ying and Greep (1972) had likewise reported advanced ovulation after EB injection on D-1.

Strain Differences in the Ovulation Response to Estrogen

Attempts to advance ovulation by administering estrogen during diestrus are not always successful. In 5-day cyclic rats of the Charles River CDF strain, advanced ovulation occurred in only 2 of 5 animals that received 9 to 15 μg EB on D-2 (Everett, unpublished). Among 12 five-day cyclic CD rats that received 15 μg EB at 1200h-1244h on D-2 (Everett and Tyrey 1982a), only five presented full ovulation two days later. Five others showed partial effects ranging from a single tubal ovum and numerous lutein cysts to a single follicle with luteinized patches. Two of the 12 rats failed to give even histological evidence of follicle activation. Cyclic rats of the DA strain had shown similar refractoriness to estrogen (Everett 1948). Of nine young 5-day cyclic DA rats treated with estrogen on D-2, seven were given 50 μg EB, one received 25 μg EB, and one was implanted with a crystal of estradiol-17β subcutaneously. Only two of the nine rats had tubal ova two days later, while the others had large Graafian follicles that were apparently unaffected. Similar refractoriness to estrogen was seen in DA SPE rats made cyclic by progesterone therapy (Fig. 19). Progesterone was first given to interrupt SPE and again on the day of proestrus to produce a short cycle, as illustrated. Six of the 10 rats then received 1.5 mg progesterone on D-1 of the second diestrus. All 10 rats were given 50 μg EB on D-2. Two days later not one rat had ovulated or had evidence of follicle activation. By contrast, when 10 similar SPE rats were injected daily with 1.5 mg progesterone for 6 days to produce artificial pseudopregnancy, injection of 50 μg EB on day 5 resulted within 48 hours in ovulation of 6 to 9 ova in 4 rats and partial luteinization of follicles in 3 rats (Fig. 19C). Here the cooperative action of progesterone with estrogen was clearly manifest.

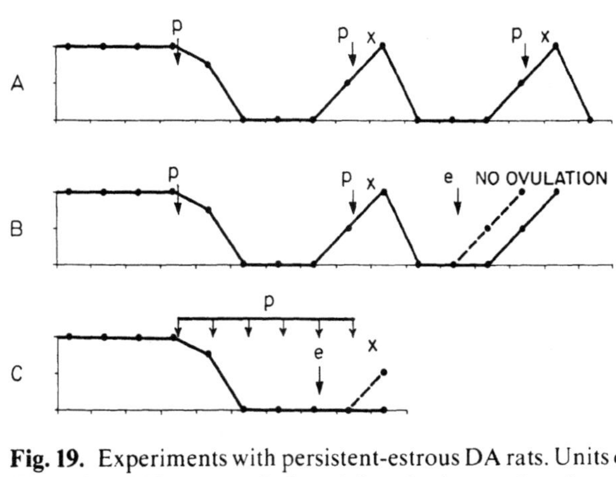

Fig. 19. Experiments with persistent-estrous DA rats. Units of the ordinate and abscissa as in Figs. 10 and 18. x = ovulation. **A** Standard procedure for producing "progesterone cycles". Each dose of progesterone (p) was 1.0 mg. New corpora formed in 70% of the cycles. **B** A progesterone cycle followed by 50 μg estradiol benzoate (e) on D-2 of the second cycle. No evidence of ovulation 48 h after the estrogen. **C** An artificial pseudopregnancy maintained by daily injection of 1.5 mg progesterone. Estrogen (e) induced ovulation or at least luteinization of follicles in most cases. (From Everett 1948)

Timing the Preovulatory Surge of Gonadotropin Secretion

Boling et al. (1941) correlated ovulation time in rats with the interval from the beginning of heat behavior. Inasmuch as the beginning of heat in these rodents is itself related to the time of day (Hemmingsen and Krarup 1937), it follows that ovulation is governed in some manner by events related to circadian periodicity. It is now common knowledge that a major determinant is a pronounced surge of LH secretion on the afternoon of proestrus, beginning during a so-called "critical period" of a few hours, cued in turn by the photoperiod to which the animal is exposed. Awareness of the close relationship of ovulation time to clock hours in rats came from the studies with progesterone and estrogen outlined in the previous section.

Whenever the newly ovulated ovaries were examined histologically on the morning afterward, the degree of development of the fresh corpora lutea was essentially identical, whether they were formed in the normal 4-day or 5-day cycle or after advanced ovulation induced by progesterone or estrogen (Fig. 24). More direct information was obtained from 28 four-day cyclic O-M rats subjected to terminal laparotomy between 0045h and 0400h early in the morning following the day of proestrus. Ovulation in progress was evident as early as 0110h, and complete ovulation was the rule after 0230h. It was thus reasonable to propose that the acute process of preovulatory follicle maturation ("swelling") had been triggered by gonadotropin secreted several hours before. Boling et al. (1941) reported that the interval from the beginning of heat behavior to ovulation was from 6½ to 10 hours long. That was suggestively similar to the interval of 10–11 hours described for the rabbit from the coital stimulus to ovulation (Waterman 1943). In that species, Sawyer et al. (1947, 1949) were able to block the reflexly induced ovulation by injecting Dibenamine (N, N-dibenzyl-β-chloroethyl amine hydrochloride, an α-adrenergic blocker) or atropine sulphate (a cholinergic blocker) soon after copulation. With that in mind, Sawyer and I postulated that these agents might also serve in rats to block a presumptive steroid-induced neural stimulus responsible for preovulatory gonadotropin discharge from the hypophysis. At our disposal were several means for testing this, since ovulation could be predicted under the following circumstances: spontaneous occurrence in the 4-day or 5-day estrous cycle; 24-hour advancement of ovulation by progesterone administered on D-3; advancement by estrogen given on D-2; and induction by estrogen during pregnancy or pseudopregnancy.

It was first necessary to determine dosages and routes of administration for Dibenamine and atropine sulphate appropriate for rats. The intravenous route was selected for Dibenamine: after a number of trials the dosage of 30 mg/kg body wt

(10 mg/ml in Ringer-Locke solution) was adopted as standard (Sawyer, Everett and Markee 1949). Trials with atropine sulphate indicated that the intravenous injection of the maximal tolerated dose (50 mg/kg) failed to block estrogen-induced ovulation in pregnant rats, perhaps because its action is too short. Consequently we chose the subcutaneous route and a dosage of 700 mg/kg (70 mg/ml Ringer-Locke solution), based on findings by Holck (1942).

Blockade of Estrogen-Induced Ovulation

These experiments were first carried out in 47 pregnant O-M females, mated with males of known fertility (Sawyer et al. 1949). The day of estrus when the vaginal smear contained spermatozoa was designated day zero.

Eight control rats injected with 50 μg EB on day 4 all had tubal ova on day 6 with corresponding numbers of fresh corpora lutea and rich accumulation of cholesterol in the corpora lutea of pregnancy (Fig. 20A). When Dibenamine was injected shortly before the administration of estrogen on day 4, ovulation failed to occur in 6 of 12 rats. Luteal cholesterol deposition was also prevented in 3 of the six (Fig. 20B). When Dibenamine was given at 0800h on the day after the estrogen injection (i.e., on day 5), it was somewhat more effective, blocking ovulation in 9 of 12 rats and preventing cholesterol accumulation in corpora lutea of 6 rats. Administration of atropine sulphate at 0900h on day 5, about 20 hours after injection of EB on day 4, blocked both ovulation and corpus luteum cholesterol deposition in 4 of 5 rats (Fig. 21). The exceptional rat produced only one tubal ovum. After LH administration on pregnancy day 5, neither Dibenamine nor atropine prevented ovulation and cholesterol storage (Fig. 21B,C). Both drugs, therefore, prevented release of LH, not its action on the ovaries.

In similar experiments (Everett & Nichols 1968; Nichols 1969), 18 O-M rats were made pseudopregnant by electrical stimulation of the cervix on the morning of estrus (day zero) at approximately 0900h. Six control rats all presented full sets of tubal ova two days after receiving 10 μg EB on day 4 or day 5. Other rats were given the standard blocking dose of atropine sulphate at approximately 1340h (6 rats) or 1540 (6 rats) on the day following the estrogen treatment. On the next morning all of the former had been blocked by the atropine. None of the latter group was completely blocked, five had ovulated completely and the other had two ruptured follicles among several that had been partially activated. The induced release of the ovulatory quota of gonadotropin had evidently occurred some 24 hours after administration of the estrogen, usually within a critical 2-hour interval.

In cyclic rats, the advancement of ovulation by estrogen administration was likewise subject to blockade with atropine (Everett and Nichols 1968; Nichols 1969). That study was carried out in 85 O-M 5-day cyclic rats, all receiving 10 μg EB on D-2. Data summarized in Table 13 demonstrated that atropine injection just before 1400h on D-3 blocked the ovulation stimulus in all but one rat, while administration at 1540h was followed by ovulation in most subjects. The critical time on D-3 for administering the blocking agent was essentially alike whether the rats had received EB on D-2 at either 0820h or 1220h, hence relatively independent of the hour of EB

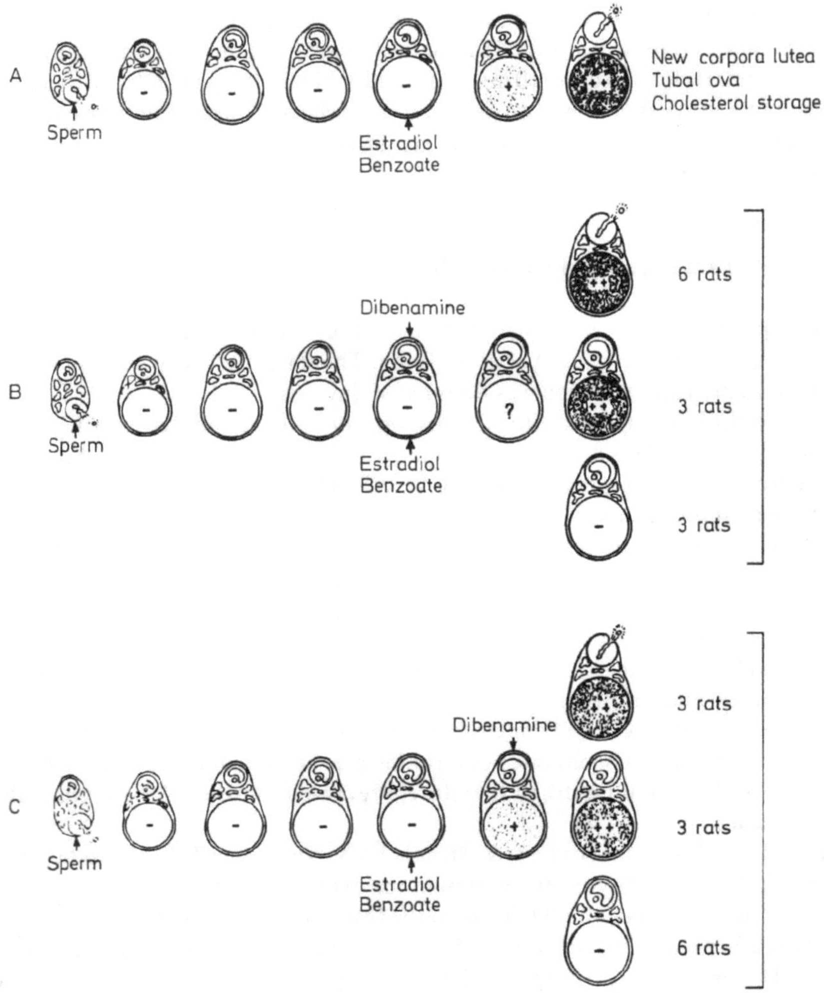

Fig. 20. Blocking of estrogen-induced ovulation and cholesterol accumulation in corpora lutea of pregnant O-M rats by Dibenamine administration. **A** Standard effects of estrogen injection of Day 4 (From Everett 1947). **B** Dibenamine injected on Day 4 just before the estrogen. **C** Dibenamine, injected 20 h after the estrogen, blocked ovulation in 9/12 rats and cholesterol accumulation in 6/12 rats. (From Saywer et al. 1949)

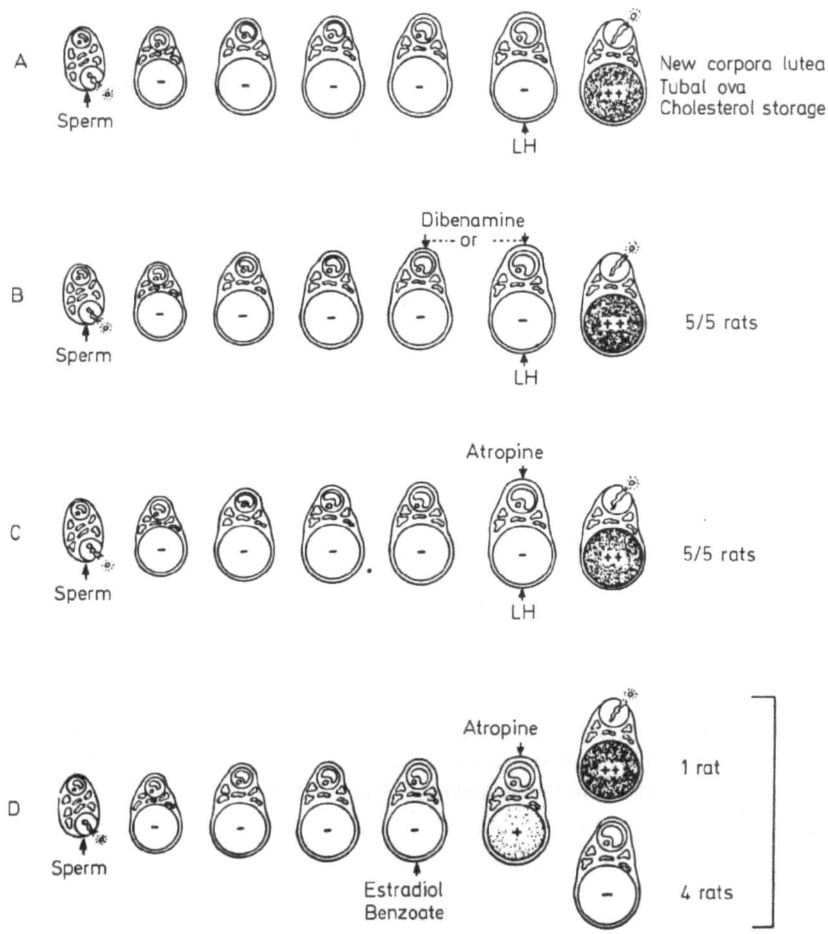

Fig. 21. **A** Extrinsic LH administered to pregnant rats induced ovulation and cholesterol deposition in the corpora lutea within 24 h after injection (Everett 1947). **B** Dibenamine injected 24 h before or simultaneously with LH, failed to block its effects. **C** Atropine sulphate likewise failed to inhibit the effects of LH. **D** Atropine injection 20 h after estradiol benzoate blocked LH release in 4/5 rats. (From Sawyer et al. 1949)

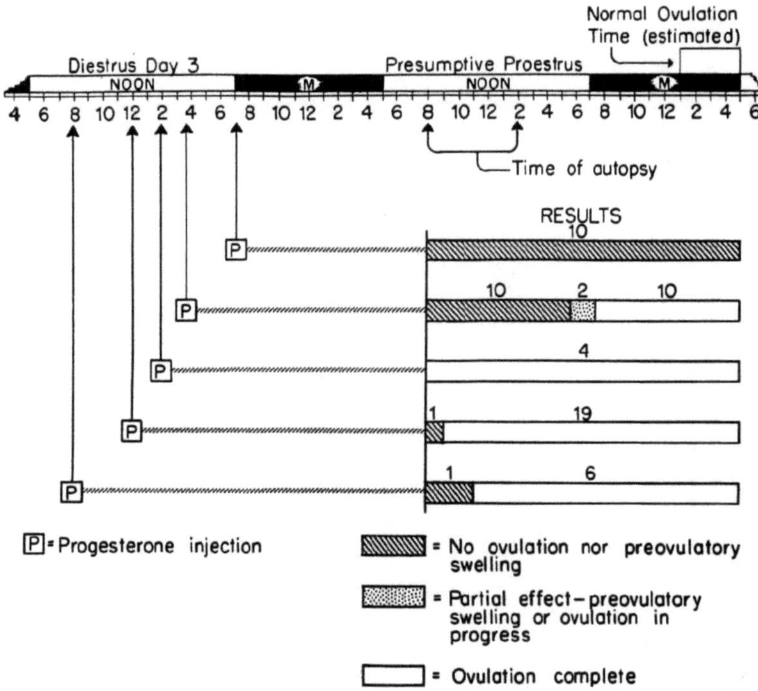

Fig. 22. Relative effectiveness of progesterone injection at different hours on diestrus day 3 in 5-day cyclic O-M rats. Lowered response to injection at 1500–1600h; no response when delayed until 1830h. Numerals over segments of bars show numbers of rats. (From Everett and Sawyer 1949)

Table 13. Effects of atropine sulfate on estrogen-advanced ovulation in 5-day cyclic O-M rats. (Modified from Nichols 1969)

Time of estradiol benzoate injection on D-2	Time of atropine injection on D-3	Total number of rats	Complete blockade	Partial effects	Not blocked	Percent complete blockade
1220 h	1345 h	7	7	0	0	100%
	1445 h	7	5	0	2	71.4 %
	1540 h	18	5	1	12	27.8 %
	1730 h	5	1	0	4	
	1830 h	6	0	3	3	6.7 %
	1930 h	6	0	0	6	
0820 h	1345 h	9	8	0	1	88.9 %
	1540 h	21	5	2	14	23.8 %

Table 14A. Effectiveness of Nembutal on day 1-2 in blocking advanced ovulation in EB-treated 4-day cycling CD rats. (From Krey & Everett 1973)

EB treatment 20 μg	Nembutal treatment 31.5 mg/kg	Results: Ovulation frequency day 3		Results: Luteinization frequency day 3
1230 Day 1	–	8/10	(5.1 ova)[a]	10/10
1230 Day 1	1300 Day 1	5/6	(8.2 ova)	5/6
1230 Day 1	1300 Day 2	0/6	–	0/6
1230 Day 1	1800 Day 2	2/6	(5.5 ova)	3/6
1500 Day 1	–	6/10	(7.7 ova)	6/10
1500 Day 1	1300 Day 2	0/6	–	0/6
1500 Day 1	1800 Day 2	3/6	(3.3 ova)	3/6

[a] Mean ovum count/ovulating rat.

Table 14B. Effectiveness of Nembutal on day 2-3 in blocking advanced ovulation in EB-treated 5-day cycling CD rats. (From Krey & Everett 1973).

EB treatment 15 μg	Nembutal treatment 31.5 mg/kg	Results: Ovulation frequency day 4		Results: Luteinization frequency day 4
1200 Day 2	–	8/9	(10.6 ova)[a]	8/9
1200 Day 2	1300 Day 2	6/6	(10.9 ova)	6/6
1200 Day 2	1245 Day 3	0/7		0/7
1200 Day 2	1500 Day 3	1/7	(4.0 ova)	1/7
1200 Day 2	1700 Day 3	1/7	(2.0 ova)	2/7
1200 Day 2	1900 Day 3	6/7	(7.2 ova)	6/7

[a] Mean ovum count/ovulating rat.

treatment. Krey and Everett (1973) confirmed this in CD rats with pentobarbital as blocking agent. Also, EB injection on D-1 in 4-day rats advanced ovulation and that could be prevented by pentobarbital given at 1300h or later on D-2 (Table 14).

Blockade of Progesterone-Induced Ovulation

It seemed probable that induction of ovulation by progesterone could also be blocked by Dibenamine or atropine. This was tested by Everett and Sawyer (1949). Seven 5-day cyclic O-M females on D-3 were given 1.0 mg progesterone at approximately 1200h and the standard blocking dose of Dibenamine 3–8 min later. None ovulated by the next morning, although there was some partial ovarian activation: 5 swollen follicles in one rat, 1 or 2 follicles in early swelling in 4 rats, no activation in 2 rats. In 3 other rats Dibenamine was injected at 0800h-0830h, followed by progesterone at noon. None of the three ovulated, but follicles were partially activated in 2 rats. Preliminary trial also showed that atropine would similarly antagonize the stimulative action of progesterone. In further study we were especially concerned with questions of timing: Does activation of the AP in response to progesterone take place at a uniform interval after injection of the steroid or is it confined to certain hours of the day irrespective of the injection time?

Fig. 22 shows the comparative effectiveness of progesterone administration at different hours. Of 20 rats injected between 0800h and 1400h nearly all had ovulated within 24 hours. Injection at 1500h-1600h ovulated only 10/22 rats, two others showing partial ovarian activation; the remainder gave no sign of activation. Later injection was uniformly ineffective in 10 rats. Hence, late in the afternoon of D-3 the 5-day cyclic rat becomes refractory to progesterone.

Treatment with progesterone early or late on D-3 followed by atropine at different intervals gave firm evidence that stimulation of the AP in response to progesterone is a function of the time of day and not of the time of progesterone administration (Fig. 23). Progesterone given at 0800h, followed by atropine at 1400h, resulted in complete failure of ovarian activation in 7/9 rats only partial activation in the others. Progesterone at 1200-1300h, followed by atropine 4 hours later, ovulated only 11/23 rats. Whereas progesterone at 1400h would be expected uniformly to produce

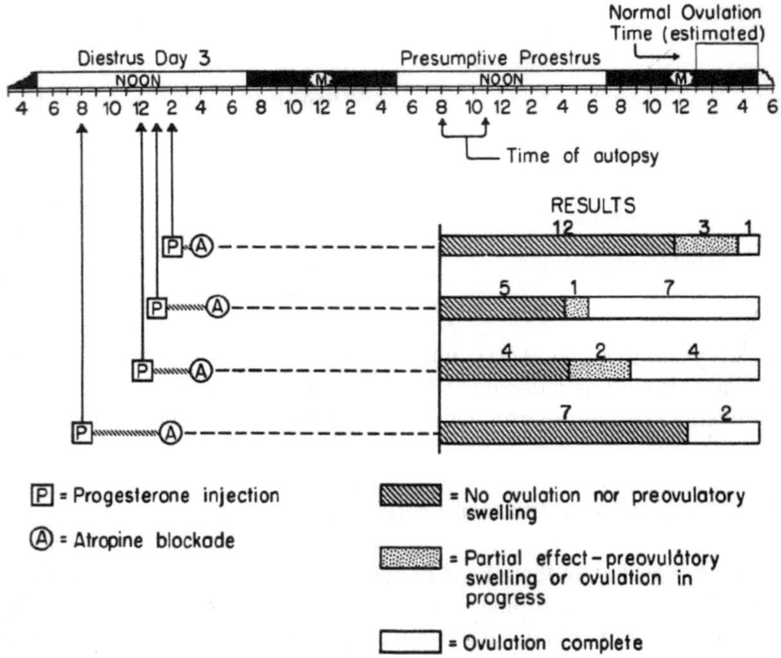

Fig. 23. Results of atropine administration (**A**) at various intervals after progesterone injection at different hours of diestrus day 3. (From Everett and Sawyer 1949)

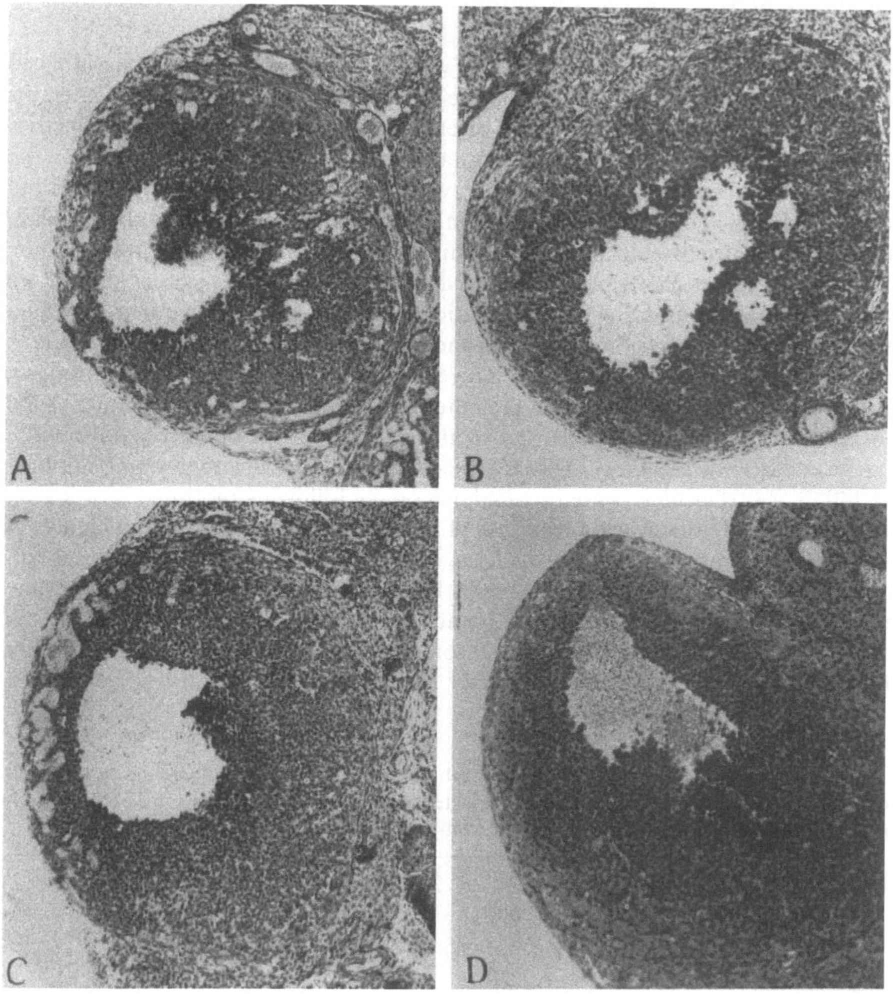

Fig. 24. New corpora lutea in 4-day cyclic O-M rats at 0800h (**A**) and 1030h (**B**) on the morning after spontaneous ovulation. **C** New corpus luteum 24h after injection of progesterone at 0800h on diestrus day 3 of a 5-day cycle. **D** New corpus luteum at 1030h on the morning after injection of progesterone at 1400h on diestrus day 3. (From Everett and Sawyer 1949)

ovulation, this was prevented by atropine administration at 1600h in 12/16 rats and only one rat showed full ovulation. Thus, the positive AP response to progesterone administration was confined to certain mid-afternoon hours on D-3 independent of the time of injection. Pointing in the same direction was the close similarity of histologic structure of the newly formed corpora lutea induced by progesterone and those obtained at similar hours after spontaneous ovulation (Fig. 24).

Blockade of Spontaneous Ovulation

Concurrently with the 1949 experiments with estrogen- and progesterone-induced ovulation, Dibenamine and atropine were used in proestrous cycling rats in attempts to prevent spontaneous ovulation (Everett et al. 1949). Both drugs were effective. Since results of preliminary trials in 5-day cyclic rats were somewhat erratic, the definitive study was limited to the 4-day cycle. The approximate injection times and results are depicted in Table 15. Following Dibenamine treatment at about 1400h, ovulation was usually either blocked or impeded. On the morning of estrus 10 of the 13 rats had no tubal ova, and histological examination of the ovaries showed no evidence of preovulatory follicle maturation in 7 of the 10 rats. The others exhibited various degrees of partial luteinization and, in one case, a single ruptured follicle as well. Earlier Dibenamine injection between 0800h and 0930h on the morning of proestrus gave similar results. On the other hand, injection at 1600h or later failed in most cases to interfere with the ovulatory process. Atropine provided more uniform blocking capacity (Table 15). In all 12 rats injected with atropine sulphate at 1400h or earlier there were no tubal ova on the following morning; histological examination of the ovaries from 10 of the animals presented no evidence of preovulatory maturation of the follicles. Injection at 1600h, by contrast, was followed by full ovulation during the night. This was convincing evidence that: 1) the normal spontaneous discharge of gonadotropin essential for ovulation occurs during the proestrus afternoon; and 2) some process subject to blockade by either Dibenamine or atropine is essential for stimulating the pituitary to release the

Table 15. Relative effectiveness of Dibenamine and atropine in blocking spontaneous ovulatory activation of the ovaries when injected on the day of proestrus in 4-day cyclic O-M rats. (Modified from Everett and Sawyer 1949)

Drug	Time Injected	No. of rats	Results[c]		
			Complete blockage	Partial blockage	Full ovulation
Dibenamine[a]	0800h to 0900h	13	7	4	2
	1400h	13[d]	7	6	0
	1600h	10	0	4	6
	1700h	3	0	0	3
	1800h	4	0	0	4
Atropine[b]	0800h	5	5	0	0
	1400h	7	7	0	0
	1600h	6	0	0	6

[a] Dibenamine hydrochloride, 30 mg/kg body weight, iv.
[b] Atropine sulphate, 700 mg/kg body weight, sc.
[c] Autopsies on the morning after the drug injection.
[d] Three additional rats autopsied late in the afternoon of estrus showed full ovulation. Supposedly delayed only.

hormone(s), since in the rabbit neither drug impairs the gland itself (Sawyer et al. 1948).

By analogy with the rabbit, in which the two drugs blocked the coital ovulation reflex, we concluded that in rats a spontaneous stimulus passes from the hypothalamus to the anterior pituitary gland at some time during a *critical period* on the afternoon of proestrus (Fig. 31). The resulting discharge of gonadotropic hormone(s) is then responsible for the rapid follicle maturation leading to ovulation 10 to 12 hours later. Here was the first evidence, although indirect, for the proestrous surges of LH and FSH that are now well recognized by virtue of the radioimmunoassay and related techniques. Furthermore, here was the first evidence for participation of a neurogenous signal in control of spontaneous ovulation. Although at that time we had no direct knowledge that atropine will pass the blood-brain barrier, Sawyer et al. (1955) later demonstrated that the ovulation-blocking doses of this and certain other drugs exert measurable depressant effects on thresholds for electrical stimulation of the midbrain reticular formation.

The list of centrally active drugs capable of blocking ovulation in rats has grown considerably. Several are α-adrenergic blockers like Dibenamine and others are anti-cholinergic, having actions similar to atropine. Several barbiturates are effective. Other notable additions to the list are morphine, chlorpromazine, reserpine (Barraclough and Sawyer 1955, 1957), urethane (Lincoln and Kelly 1972), and cannabinoids (Cordova et al. 1980; Nir et al. 1973). Pharmacologic approaches for study of the central neural control of pituitary functions have recently been very fruitful.

Pharmacologic Intervention Compared with Hypophysectomy During the Critical Period

The two-hour "critical period" delineated by use of Dibenamine and atropine was much longer than the time required in the rabbit for discharge of an ovulating quantum of gonadotropin (1 hour: Fee and Parkes 1929; Smith and White 1931; Westman and Jacobsohn 1936). Furthermore, in rabbits, that phase of the reflex stimulus which is sensitive to atropine and Dibenamine is completed within a very few minutes (Sawyer et al. 1949). To estimate in rats when the spontaneous neurogenous stimulus begins in different individuals, its duration, and the time needed for discharge of the requisite amount of gonadotropins, blocking doses of atropine sulphate were administered to proestrous cyclic rats at various times during the critical period. In parallel experiments, rats were hypophysectomized at times similar to those in some of the atropine treatments (Fig. 25).

The atropine experiments were performed in two series (Everett and Sawyer 1953; Everett 1956b). Altogether, 100 proestrous cyclic O-M rats, selected for histories of at least two regular 4-day cycles, were each injected with atropine sulphate at the approximate times indicated in Fig. 25. Parapharyngeal hypophysectomy (Everett 1956b) was carried out in 35 similar rats near one of the three times indicated. The technique was standardized so that the gland was removed within 7–10 minutes after ether anesthesia began. The effective blocking time was arbitrarily set at one minute

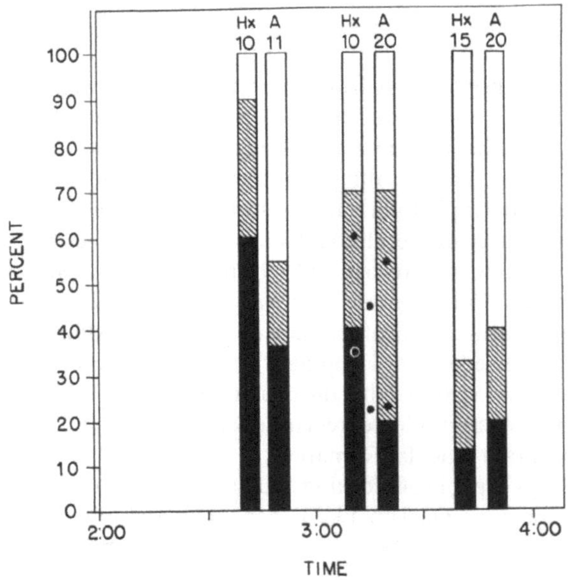

Fig. 25. Results of hypophysectomy (Hx) or injection of atropine (**A**) at different times during the proestrus critical period in 4-day cyclic O-M rats. *Solid black:* complete blockade. *Cross hatching:* partial blockade. *White:* complete ovulation. The number of rats in each time-group is shown above. The vertical pairs of dots mark the averages for all time groups for Hx and A, respectively. The second pair (between columns) are derived from the present series and the previous 1953 results. The lower dot in each pair registers percent totally blocked; the vertical spacing between dots registers percent partially blocked. (From Everett 1956)

after introduction of ether, since control experiments had indicated that ether itself can temporarily delay gonadotropin discharge. Blake (1974), by radioimmunoassay has documented such delay of LH release by ether. Completeness of hypophysectomy was determined by examining the cranial floor after the experiment.

The effects of either atropine injection or hypophysectomy were evaluated at terminal laparotomy on the next morning between 0800 and 1000h. The oviducts were searched for ova and the ovaries were examined under a dissecting microscope for fresh corpora lutea. In the absence of tubal ova, the ovaries were preserved for histologic study of serial sections. The respective criteria for complete blockade, partial blockade, and full ovulation were as follows: To represent *complete blockade*, the ovaries must lack all traces of luteinization of the current set of Graafian follicles, there must be no evidence of secondary liquor, and the germinal vesicles must be intact in all oocytes. The presence of 7 or more tubal ova and/or fresh corpora lutea was arbitrarily regarded as evidence of *full ovulation*, since 7 was the smallest number of offspring in our breeding of O-M rats. The category of *partial blockade* ranged from cases in which fewer than 7 follicles ovulated and luteinized, at the one extreme, to others in which there was only minimal histological evidence of prelutein change, at the other extreme (see Fig. 26).

Both atropine injection and hypophysectomy blocked progressively fewer rats as the time of intervention progressed (Fig. 25). Partially blocked rats were encountered in all groups. The overall percentages of partially blocked rats in the atropine-treated and the hypophysectomized groups were almost identical, 25.0% and 25.7%, respectively. Since the chance of interrupting a process is proportional to its duration, it was possible to estimate the requisite length of stimulation and the time required to release the minimal amount of hormone for complete ovulation: roughly half an hour for both processes. This was interpreted to mean that both the atropine-sensitive neurogenous stimulus and the actual release of hormone proceed more or less in parallel. This interpretation was based on the assumption that these inbred animals were essentially alike in individual requirements for gonadotropic stimulation, together with evidence that the onset of blocking action of atropine in the massive doses is abrupt soon after injection (Sawyer et al. 1955), and that its action is essentially all-or-none (Everett and Sawyer 1953).

Obviously some delay must nonetheless occur between the beginning of the neural phase of the stimulation process and the appearance of LH in circulation.

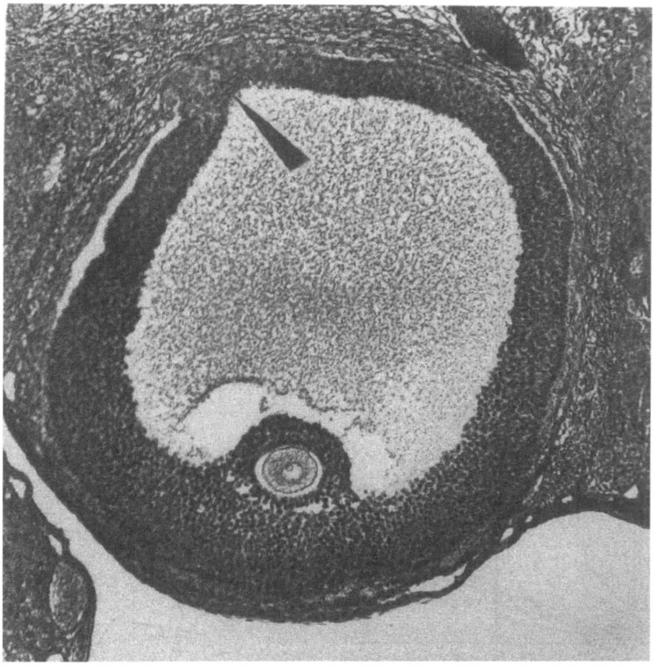

Fig. 26. Localized lutein plaque (arrow) in the wall of a Graafian follicle otherwise unaffected by luteinizing hormone, indicative of almost complete blockage of the proestrus LH surge. (From Everett 1952)

Time must be consumed first of all in the strictly neural apparatus, then in the discharge of LHRH into the pituitary portal vessels, in the activation of the gonadotropic cells, and in the discharge of sufficient hormone to produce some recognizable effect on the ovary. Whether or not there is a pronounced delay overall, it seems apparent that the amount of gonadotropin released is proportional to the duration of the stimulus. If the stimulus is cut short the amount of LH released is diminished.

Fig. 27 demonstrates that in individual rats of the population examined, both the neurogenous stimulus and the LH surge began at different times in different individuals. Such variations became widely recognized once the radioimmunoassay techniques came into general use (Everett et al. 1973; Blake 1976, Schuiling et al. 1976). It has also been well established that an excess of LH is normally released during the surge, usually far above the minimal amount needed for ovulation. Indirect evidence of that was seen in the hypophysectomy experiments just described (Everett 1956b). The minimal amount (the ovulation quota) has been estimated by Grieg and Weisz (1973) as about 15% of the full surge.

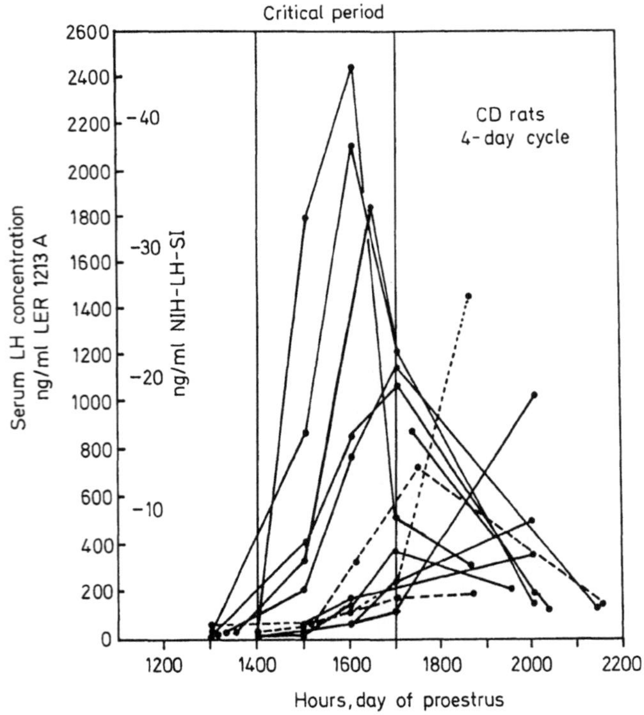

Fig. 27. Varied times of onset of spontaneous preovulatory surges of LH in 12 proestrous 4-day cyclic CD rats. The limits shown for the critical period in CD rats were somewhat longer than those defined earlier for O-M rats. (From Everett et al. 1973)

Effect of the Lighting Rhythm on the "Critical Period"

The plan of this experiment (Everett unpublished) was simply to advance the time-switch in the animal room by 3 hours on a given day, and thereafter to test different groups of proestrous 4-day cyclic rats on successive days by injecting atropine at 1400h of the old time-scale. As before, the standard atropine sulphate dose was 700 mg/kg body wt., sc. Each rat was killed for examination on the morning after injection. Fig. 28 shows that during the first 6 days after the time-shift, there was steady decline in the proportionate numbers of animals blocked. The combined results for days 5 and 6 were very similar to those normally found after injection of atropine at 1500h, midway during the critical period. Hence, they represented an adjustment of the animals themselves amounting to about 1 hour. On day 12, the hour of atropine injection was changed to 1200h on the normal time-scale, now 1500h on the *experimental* time-scale (Fig. 29). All 10 rats failed to ovulate, 3 of the 10 presenting swollen follicles. Had there been full adjustment to the new lighting schedule, several should have ovulated in spite of the atropine treatment (see Fig. 25).

The time-switch was returned to standard setting 13 days after the original shift, and other groups of rats were tested with atropine. Ten rats were injected at 1400h with atropine sulphate during proestrus on either the 13th, 14th or 15th day after the return to standard lighting. All were fully blocked. Thus, the readjustment to the normal schedule of lighting was at least as rapid as in the reverse direction.

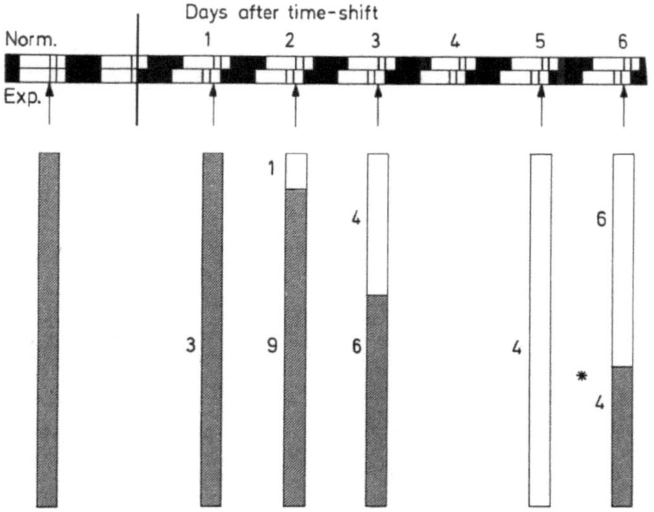

Fig. 28. Progressive downward shift of blocking effectiveness of atropine injected at 1400h of proestrus during days after an abrupt 3-hour shift of the lighting schedule (4-day cyclic O-M rats). The respective normal and experimental periods of light and dark are shown above, timing of the atropine injection by the arrows, and the numbers of rats failing to ovulate by the hatched segments of the bars. The asterisk marks the average of those blocked by atropine on days 5 and 6. (From Everett 1970)

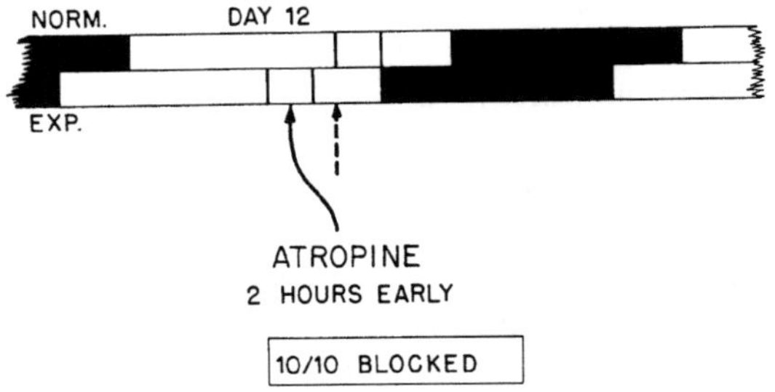

Fig. 29. Same experiment as in Fig. 28. Atropine, injected on day 12 in the middle of the presumptive new critical period, blocked ovulation in all 10 rats although 3 rats had follicles in preovulatory maturation. Adjustment to the new lighting schedule judged to be not yet complete. (From Everett 1970)

From these results it is apparent that the influence of the lighting rhythm is not an immediate one; full adjustment to a new rhythm seems to require 3 weeks or longer. Data of Baldwin and Sawyer (1979) suggest that by 3 weeks, shifting of the LH surge may be almost complete. This is the same order of time as that needed for adjustment of the activity rhythm and sexual behavior in this species following a time-shift of 12 hours (Hemmingsen and Krarup 1937). One may propose that the day-to-day rates of adjustment of both the activity rhythm and the ovulation rhythm may be proportional to the shifting differences between the old and new environmental rhythms.

Circadian Periodicity in Mechanisms Governing the Ovulatory Surge of Gonadotropins

The view that a 24-hour periodicity exists in the control mechanisms for the ovulatory discharge of gonadotropins by the rat hypophysis was first outlined by Everett and Sawyer (1949) in discussing the 24-hour advancement of ovulation by administration of progesterone and the fact this effect could be blocked by Dibenamine or atropine. Noting that the cyclic element in the control apparatus probably "resides in, or operates though, the anterior hypothalamic area" we judged that "one of its functional attributes (in the rat at least) is an intrinsic 24-hour rhythm." We entertained three alternative hypotheses about the process by which progesterone and estrogen may act in promoting ovulation, all based on the assumption of circadian rhythmicity of that rostral hypothalamic element. The first hypothesis proposed that neurohumoral stimulation passes to the hypophysis each day and that response of the gland depends on its sensitization by the steroids. The second hypothesis was that the principal site(s) of action of the steroids is in the tuberal hypothalamus or median eminence. Given adequate steroid influence there, the daily signal from the rostral timer invokes the neurohumoral stimulus. According to the third hypothesis the rostral apparatus has a diurnal rhythm of sensitivity but does not discharge until activated by the sex steroids; this does not deny important steroid influence at the other sites.

In relation to these findings we cited preliminary experiments with pentobarbital sodium as blocking agent for spontaneous ovulation. Various amounts from 27 to 45 mg/kg body weight were given intraperitoneally to proestrous 4-day cyclic O-M rats shortly before 1400h. Ovulation was consistently blocked that night, but if such rats were kept another day they ovulated during the second night. Further investigation disclosed a renewed critical period on the second afternoon, defined by the use of atropine (Everett and Sawyer 1950). Pentobarbital treatment on that second day beginning at 1400h could also prevent ovulation, provided that an additional injection was given 2 hours later or the initial dosage at 1400h was raised to 45 mg/kg. A further postponement for 24 hours followed, with another critical period developing on the third afternoon (Fig. 30). Blockade with pentobarbital at that time resulted in a follicular cycle without luteinization; otherwise ovulation occurred that third night. Control experiments established that the 24-hour delays were not due to residual, prolonged blocking action of pentobarbital, for injection at 1000h on the day of proestrus failed to block any of 4 rats and injection at 1200h failed in 2 of 6 rats.

During the follicular cycle resulting from repeated blockade day-to-day, follicle growth continued into the second day and then faltered (Fig. 32). Although there was little enlargement between day 2 and day 3, the follicles were nevertheless competent on day 3 to respond to a gonadotropin surge by ovulating. But if blocked on that day too, they became frankly atretic by the morning of day 4. In company with the continuing follicle growth evident on day 2, the uterus remained moderately distended and hyperemic, but thereafter lost these manifestations of estrogenic stimulation. The vagina remained cornified an extra day into day 3, but by day 4 had regressed to stage IV or stage V. Interestingly, that vaginal sequence was the same whether there had been blockade for one, two, or three days. If the experiments were extended, the ensuing diestrus lasted for only 2 or 3 days under any circumstance; a new proestrus then appeared and regular cycles resumed.

Butcher et al. (1975) made a detailed study of plasma concentrations of LH, FSH, prolactin, progesterone and estradiol-17β in rats blocked with pentobarbital for either 1 or 2 days, comparing them with normal rats receiving only the vehicle. Estradiol concentrations reached maximum at 1200h on the day of normal proestrus. After the first blockade, the concentrations dropped somewhat during the night, but to a lesser degree than in control rats not receiving pentobarbital. In blocked animals, estradiol rose sharply again toward noon of the second day, but when pentobarbital was repeated on that day the estradiol level became comparatively low by evening. In agreement with our observations, Butcher et al. reported that after pentobarbital blockade for 2 days, not only did the follicles remain histologically normal on day 3 and competent to ovulate, but the oocytes remained in diapause with intact germinal vesicles. However animals copulating that night presented a somewhat lowered fertilization rate and an increased tendency toward abnormal embryonic development similar to that occurring in the case of spontaneously delayed ovulation in 6-day cycles.

Circadian periodicity of phasic surges of LH in the rat has been confirmed, especially well in ovariectomized subjects treated either with repeated EB injection (Caligaris et al. 1971) or with implantation of silastic capsules containing estradiol (Banks and Freeman 1978; Legan et al. 1975; Wise et al. 1981). The absence of daily surges throughout the short cycle in rats is apparently due to periovulatory suppression by progesterone. In the ovariectomized, estrogen-primed rat, once an initial LH surge has taken place, a single injection of progesterone suppresses further surging for several days (Caligaris et al. 1971; Freeman et al. 1976).

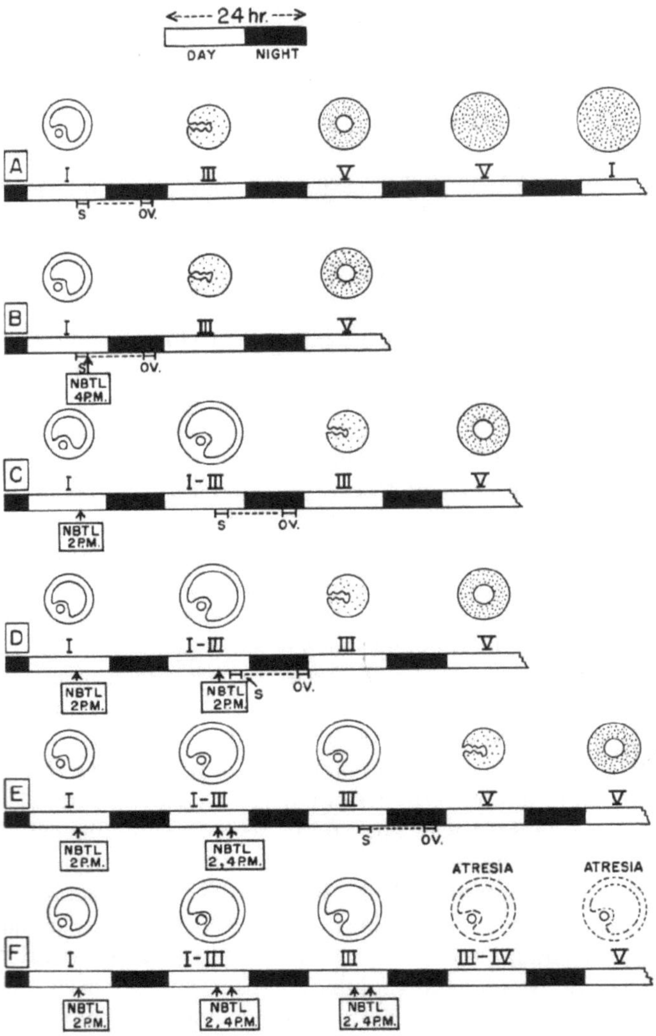

Fig. 30. Representations of the normal 4-day cycle in O-M rats (A) and characteristic results of different regimens of pentobarbital treatment (B-F). Vaginal stages are shown over each time scale while the symbols above these show corresponding follicle and corpus luteum stages. "S" marks the normal and experimental critical periods. "OV" indicates the normal ovulation time in "A" and the estimated time elsewhere. "NBTL" indicates intraperitoneal injection of pentobarbital. (From Everett and Sawyer 1950)

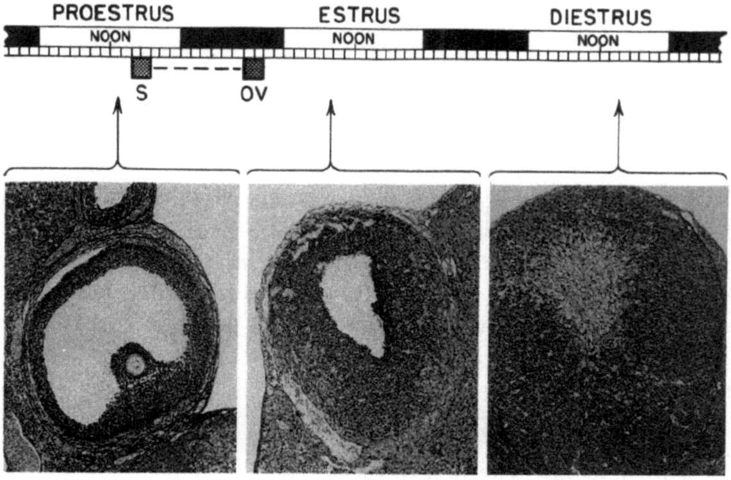

Fig. 31. Follicle and corpus luteum histology characteristic of proestrus and the next two days of the 4-day cycle (O-M rats). "S" marks the critical period for pituitary activation and "OV" the time of ovulation. (From Everett and Sawyer 1950)

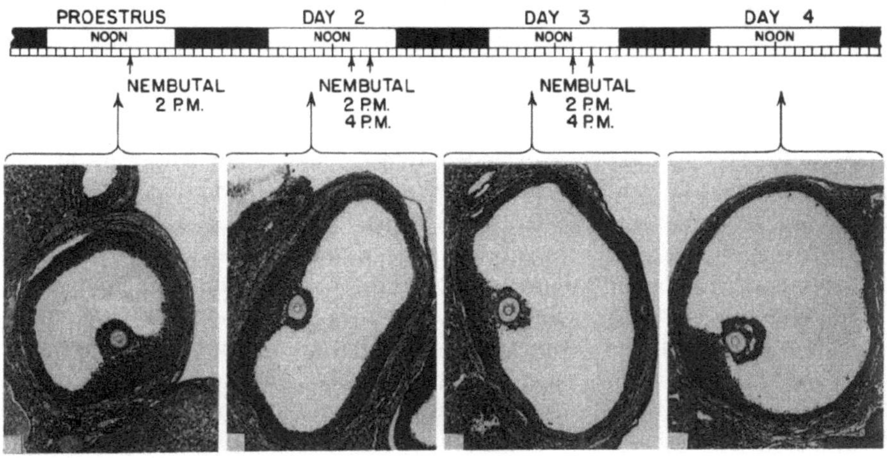

Fig. 32. Results of blocking pituitary activation for 3 successive days. Although the follicles were competent to respond to an LH surge on days 2 and 3, by day 4 they were usually atretic. (From Everett and Sawyer 1950)

There are interesting variations of the influence of illumination cues on timing of the LH surge in other rodents, notably hamsters and guinea pigs. Intact hamsters exposed to short photoperiods display daily afternoon surges of LH correlated with low levels of estrogen and high progesterone. By contrast, subjects exposed to long days and treated with estrogen show intensified surges at first, but later suppression; progesterone hastens the suppression (Bridges and Goldman 1975; Seegal and Goldman 1975). In the guinea pig, according to Terasawa et al. (1979 a,b), the timing of an estrogen-induced surge of LH in an ovariectomized animal depends partly on the timing of estrogen injection, but also partly on the time of day. The surge is larger when it occurs during the dark phase of the daily rhythm than during the light phase. Spontaneous episodic surges in ovariectomized guinea pigs tend to be more frequent during the dark phase. Daily phasic LH surges in response to estrogen priming are not seen in such animals.

The Empirical Nature of the "Critical Period"

The so-called "critical period" is not absolute in either the time of its beginning or in its duration. Although our results from atropine and Dibenamine blockade as first presented indicated a sharply limited two-hour period during which spontaneous activation of the pituitary occurs and sufficient gonadotropin is released to produce full ovulation (Everett et al. 1949), the restricted 1400h-1600h range of time was evident in only those rats having 4-day cycles. We were aware that in 5-day O-M rats the range was somewhat longer. Among 9 proestrous 5-day rats injected with atropine at 1600h, full ovulation was encountered in 3 cases, 3 rats were partially blocked, and 3 others were completely blocked. Similar results were met with in 4-day rats of the CD strain; when they were tested with pentobarbital (Everett and Tejasen 1967) the critical period extended from 1330h to 1700h. It appears that the term "critical period" can apply to only a particular set of animals under particular conditions. From that point of view, no individual rat has a critical period, but there are certain hours when and if conditions are appropriate she can begin to release the ovulatory surge of gonadotropins. The time limits will be determined not only by these conditions, but by the methods used and the end-points chosen. Blake (1974) differentiated the "critical period" during which the minimal ovulation quota of LH is normally released, from a longer "activation period" of sustained high-level LH secretion, and from a still longer "potential activation period" continuing until 2100-2200h. During this extended time, an LH surge can occur following temporary suppression by various agents having short-term blocking action (ether, ethanol, nicotine, or low doses of barbiturates or urethane). Earlier mention was made of the fact that the normal surge sometimes begins very late. There are also indications that the overall potential activation period may begin somewhat earlier than 1400h. In proestrous rats, injection of progesterone at 0900h to 1200h can induce ovulation in spite of a blocking dose of atropine given at 1400h (Everett 1951; Redmond 1968; Zeilmaker 1966). There is also evidence that surgical stress during the morning of proestrus may promote a precocious LH surge (Nequin and Schwatz 1971).

Central Neural Control of Spontaneous Ovulation: The Rat Model

The facts that both steroid-induced and spontaneous ovulation in rats can be prevented by treatment with centrally acting drugs and that the controls exhibit circadian rhythmicity implied participation of the central nervous system. Neural control was also implied indirectly by the report by Dempsey and Searles (1943) that LLPE rats formed corpora lutea after copulation. In the estrous rabbit and cat, several workers induced ovulation by electrical stimulation of the hypothalamus or the amygdala (see Harris 1972). Yet not until 1957 was there comparable direct evidence of a role of the brain in species that ovulate spontaneously. In that year Bunn and Everett reported ovulation induced in alert LLPE rats by electrical stimulation through electrodes chronically implanted in the amygdala or septum pellucidum. Critchlow (1957, 1958), employing cyclic rats anesthetized with pentobarbital during the proestrus critical period, succeeded in inducing ovulation by stimulation through electrodes stereotaxically placed deep in the medial hypothalamus. Since then the pentobarbital-blocked rat has become a favorite experimental subject in this and other laboratories.

Electrochemical Stimulation (ECS)

The electrodes used by Critchlow were in most cases bipolar coaxial stainless steel units, two units being inserted into the hypothalamus about 1 mm apart across the midline. The stimulus consisted of 15-sec trains of 1-msec 1.5–4.0 V monophasic rectangular pulses at 100 Hz from a Grass S-4 stimulator at 15-sec intervals for 10 min. The microamperage was not mentioned. Since the S-4 stimulator without an isolation unit delivers only positive current when adjusted for monophasic pulses and since there was no indication whether the core or the barrel of the electrode was connected as anode, one can only assume that the connection was random. Critchlow noted that "some difficulty was encountered in selecting effective parameters which did not cause d.c. lesions." In retrospect, it seems likely that small irritant lesions were produced by deposition of iron from either the electrode cores or barrels and that the stimuli were therefore electrochemical and not electrical in the conventional sense. It is also possible that in the region of the arcuate nuclei some of the effective "stimulations" were the result of tissue damage *per se* (see pp. 66–69).

Attempting to repeat Critchlow's findings we tried a variety of electrical parameters and electrode locations, using coaxial units like those that he described (Harp, Riley, Radford, Christian and Everett unpublished). Eventually, after frequent failures of ovulation following electrode placement in the basal tuber, we explored more rostrally and found that positive results were easily obtained from stimulation in the medial preoptic area (MPOA) or anterior hypothalamic area (AHA) (Fig. 36). A close relationship became apparent between the effectiveness of a given stimulating current and the size of the small electrolytic lesion that it produced. Application of the Prussian Blue histochemical test disclosed deposits of iron around the electrode site. That led to use of continuous direct current instead of monophasic pulses and to examination of the relative effectiveness of given amounts of current in microamperes through electrodes of stainless steel, reagent grade iron, and platinum, respectively (Everett and Radford 1961a,b). The outcome was that ovulation was induced consistently by anodic deposition of iron in the MPOA from steel or iron electrodes, but not by electrolytic lesions made with platinum. Furthermore, the positive effect was obtained by microinjection of $FeCl_3$ or by implanting minute crystals of $CuCl_2$. In one experiment, 6/7 rats ovulated after small amounts of rust were compacted and inserted unilaterally into the MPOA. Comparable amounts of ground glass were ineffective, on the other hand.

Although Critchlow stimulated his animals bilaterally and in our early trials we likewise utilized bilateral electrodes, unilateral ECS eventually proved to be sufficient. That and the fact that medial preoptic electrolysis with a platinum electrode failed to induce ovulation, assured that the effect of iron-induced lesions is an active process not due to destruction of inhibitory neurons.

Electrodes for ECS in our laboratory were essentially as described by Critchlow (1958): coaxial, comprising a 0.47 mm tube containing a 0.15 mm wire that protruded 0.3–0.5 mm. Both elements were insulated to the tip, excepting a bare 0.5 mm ring at the end of the tube and the cut end of the wire. Both tube and wire were

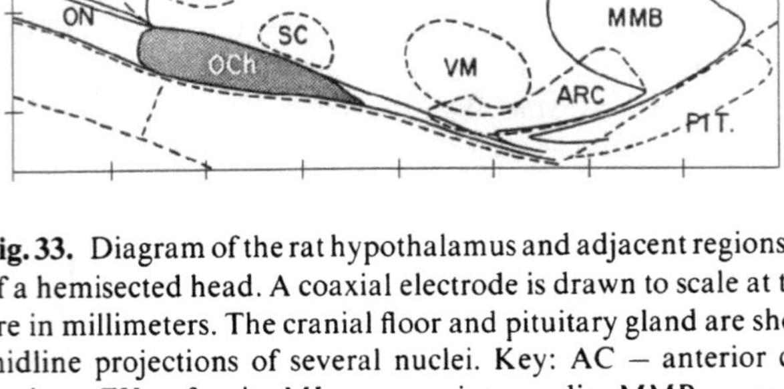

Fig. 33. Diagram of the rat hypothalamus and adjacent regions, from a camera lucida tracing of a hemisected head. A coaxial electrode is drawn to scale at the upper left. Scale markings are in millimeters. The cranial floor and pituitary gland are shown in broken lines as are the midline projections of several nuclei. Key: AC – anterior commissure, ARC – arcuate nucleus, FX – fornix, MI – massa intermedia, MMB – mammillary body, OCh – optic chiasma, ON – optic nerve, PIT – pituitary, PO – preoptic nucleus, PV – paraventricular nucleus, SC – suprachiasmatic nucleus, VM – ventromedial nucleus. Rostral limit of third ventricle indicated by the arrow. (From Everett 1961)

of stainless steel, thought to be #304 alloy. The unit was cemented to the end of a larger tube several centimeters long for firm mounting in the stereotaxic instrument. For bilateral stimulation two units were assembled on a common carrier tube. When the coaxial unit was used for monopolar stimulation, only the core alone was wired (as anode), with the tube serving merely as mechanical support; a brass rod in the rectum then served as cathode.

The stimulator was a Grass S-4C with an IU-4 isolation unit. In series with the preparation was a variable 500 KΩ resistor, usually set near its mid-point. That high resistance eliminated the capacitance artifact from the oscilloscope display (see Fig. 38). Connected across a 1000Ω resistor, the oscilloscope was calibrated for current with a precision microammeter before each day's experiments.

To evaluate electrode location, size of the lesion and the amount of iron deposited, the following procedure was routine at terminal laparotomy performed under an excess of pentobarbital. After examination of the reproductive tract the head of the animal was perfused through the heart, first with 0.9% physiological saline containing 1% potassium ferrocyanide as Prussian Blue reagent, then with 10% formalin in saline. Once the head was removed and skinned, the cranium was trimmed and placed for several days in 10% formalin containing 1% $CaCl_2$. Later the brain was appropriately dissected to leave a block of tissue containing the electrode track and lesioned area, together with adjacent structures to allow localization in histologic sections. Serial 20 μm sections, either frozen or from paraffin blocks, were stained in neutral red.

Characteristics of the electrochemical lesion and its temporal changes

Typical lesions are shown in Figs. 34 and 35 as they appear on the day after the electrolysis. The darkened zone close to the site of the electrode tip represents a heavy deposit of iron colored by Prussian Blue. This region is also marked by a considerable degree of coagulation, especially after large currents. The surrounding pallid halo contains many small cells judged to be neuroglia, but no surviving neuronal perikarya. The delicate myelinated fibers normal to the region are likewise absent in sections stained by Luxol Fast Blue or the Weil technique. Fine punctuate deposits of Prussian Blue can be recognized throughout the halo under high magnification with optimum illumination. Some vacuolation is evident in Fig. 35, representing accumulation of gas at the end of the cathodic barrel. With high current levels such accumulations greatly distort the lesion, but at the comparatively low values of current used in the examples illustrated this effect was not evident. The accumulation of gas can readily be avoided by simply using as anode a monopolar electrode (for example, the core of a coaxial unit) together with an indifferent cathode placed extracranially. A lesion thus produced is represented in Fig. 41.

Unlike the lesion as seen on the next day, the picture is significantly different soon after the passage of the lesioning current, at a time when the stimulative action is most pronounced in terms of gonadotropin secretion (Judith Furman, Linda Smith and Everett unpublished). The halo has not yet formed and for several minutes only the coagulum is to be seen, with heavy concentration of iron at its periphery. The surrounding region shows no histological evidence of damage and at first little

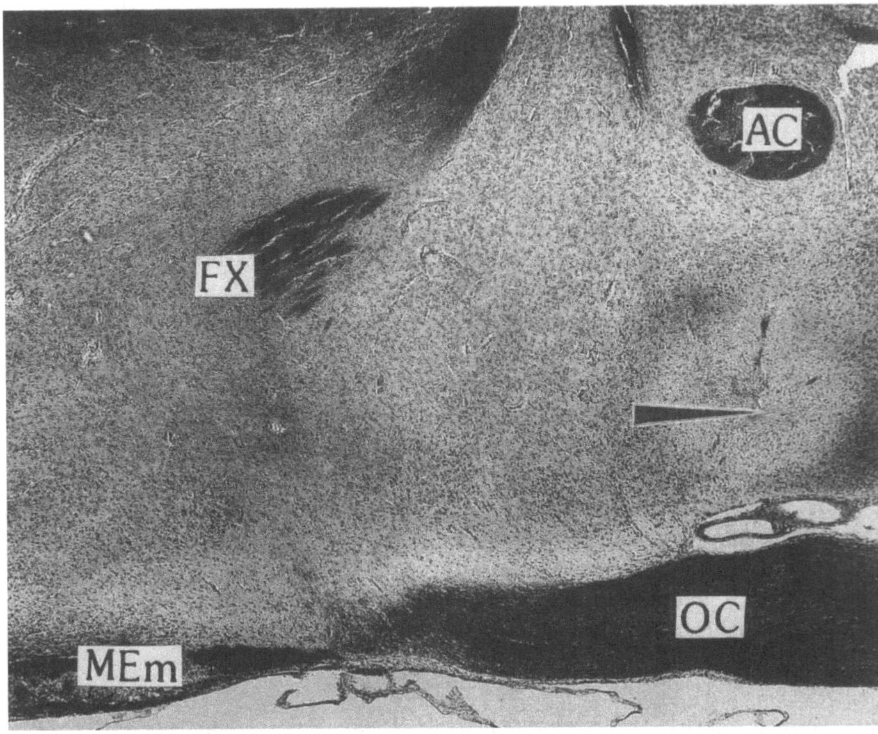

Fig. 34. A typical medial preoptic electrochemical lesion as seen on the next morning. Position of electrode tip is shown by the arrow. This irritant focus was formed by monophasic anodic rectangular pulses (1 msec, 150 μA, 100 Hz × 60 sec). Proestrous 5-day cyclic O-M rat, hypophysectomized 41 min after passage of current. Twelve tubal oval present at necropsy. Parasagittal section: AC = anterior commissure; FX = fornix; MEm = median eminence; OC = optic chiasma. (From Everett et al. 1964)

evidence of iron, save close to the coagulum. There one may find Prussian Blue within perikarya and in some fibers (axons, dendrites?) extending outward for short distances. Preparations made 30 to 60 minutes after passage of current show fine dots of the reaction product throughout the region of the eventual halo, yet no visible damage to perikarya. Iron is also demonstrable then in the walls of blood vessels near the electrode site and occasionally several hundred micrometers distant. Occasionally, Prussian Blue has appeared in neuronal processes coursing in the plane of the section, suggesting that some of the "punctate" reaction product represents such processes in cross-section. Van der Schoot and Lincoln (1978b), in brains perfused with glutaraldehyde-saline 1–3 hours after preoptic ECS, detected iron within axons at points far distant from the lesioned area. Their sections were stained by Timm's technique for intensification of heavy metals (Tyrer and Bell 1974).

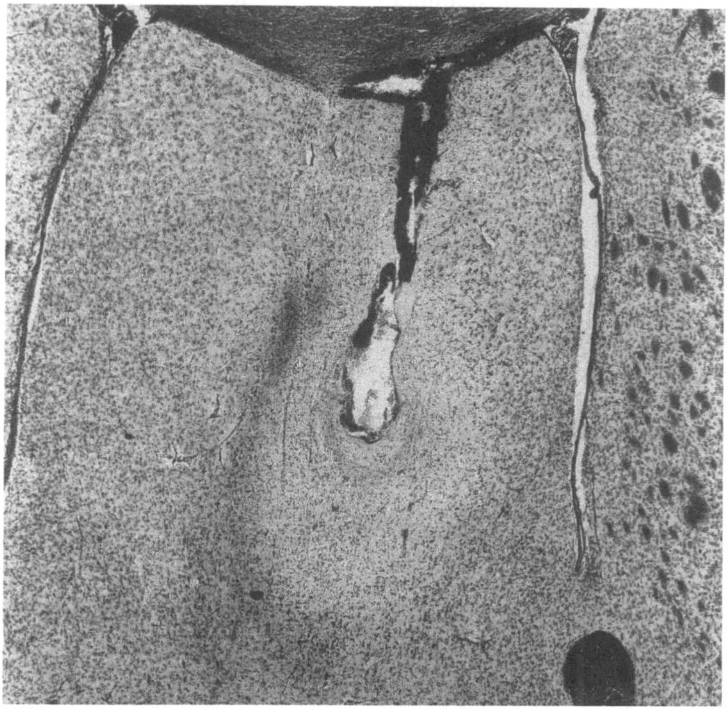

Fig. 35. A large electrochemical focus formed by 100 μA direct current for 30 sec. Proestrous 4-day cyclic O-M rat. Nine tubal ova present next morning. Coronal section through the septum. (From Everett et al. 1964)

Although the mechanism whereby an electrochemical lesion brings about stimulation was obscure for many years, there has been substantial recent progress toward an understanding. Colombo, Whitmoyer and Sawyer (1974, 1975), comparing the effects of iron deposits on multiunit activity (MUA) in various regions of the rat brain, found excitation near the lesion site although in only those areas of the brain that are rich in perikarya. It seems likely that this rise of MUA occurs in perikarya placed where the halo will form later. We had originally postulated that both cells and fibers might be activated. According to van der Schoot et al. (1978) numerous individual neurons about 1 mm caudal to a preoptic electrochemical focus displayed action potentials at rates increasing progressively from 0.5 to 15–20/sec following a delay of 10–20 min, the high rate being then sustained for recording periods of 90–230 min. Oddly, the early time of low activity corresponds to that when high levels of LHRH are said to appear in the basal hypothalamus and the pituitary portal vessels (Barr and Barraclough 1978; Eskay, Mical and Porter 1977). Inasmuch as Dyer and Burnet (1976) could find no excitatory response of identified units after electrophoretic transfer of Fe^{2+}, but only depression, they concluded that the stimulative effect is actually due to disinhibition. However, Reid and Sypert (1980) examining the firing patterns of neurons near the site of electrophoretically introduced iron, observed that within 3 mm most units were excited, while at greater distances the proportion showing increase dropped to 30% and a considerable number of units were inhibited. Colombo and Saporta (1980) demonstrated local uptake of 2-deoxyglucose after ECS or direct introduction of iron compounds into the rat brain. Further insight into the irritant action of iron in brain tissue is offered by investigations relating to post-traumatic epileptiform seizures. Reviewing pathological effects of iron, Aisen (1977) suggested that the underlying factor is the ability of the Fe^{2+}/Fe^{3+} couple to function as an oxidative catalyst that produces free radical intermediates, including the superoxide radical anion. Willmore et al. (1978–1983) likewise ascribed the irritant effect to formation of free radicals and powerful oxidants leading to neural lipid peroxidation. However, it still remains to be shown how such agents as intermediaries provoke action potentials in the periphery of an electrochemical lesion and sustain them for several hours before neuronal deterioration becomes manifest.

Electrical Stimulation (ELS)

Whereas the electrochemical method has certain advantages for activating the hypothalamo-pituitary system for LH release, especially because the irritative action persists long after the brief passage of electric current and the action involves a considerable bulk of brain tissue, ECS has undesirable features precluding its use for some purposes. The tissue damage makes it inappropriate for chronic experiments, and the duration of its action cannot be controlled or even exactly measured. On the other hand, electrical stimulation with non-polarizing electrodes of platinum and with biphasic pulse trains limits the tissue damage to the electrode track itself, allows control of the duration of stimulation, and potentially affords better localization than can be accomplished with ECS. Our methods for electrical stimulation have been as follows (Everett 1965; Everett and Tyrey 1981).

Electrodes were of two varieties. *Spaced dual bipolar assembly*: Each element was a 0.15 mm platinum wire (90% platinum + 10% iridium) insulated with Teflon and ensheathed in fully insulated stainless steel tubing for protection. The two elements were mounted 2 mm apart and the wire tips were exposed on the facing sides for approximately 1 mm. *Coaxial electrode*: This type consisted of a 0.15 mm Teflon-insulated platinum-iridium wire encased in 0.4 mm platinum tubing, the unit being insulated with Epoxylite resin. The wire core protruded 0.3 mm and was bare only at the tip; the tubing was bare at its distal end for ~ 0.2 mm (see Fig. 39).

Matched biphasic pairs of rectangular 1 msec pulses were delivered by two Grass stimulators with isolation units in series. The peak-to-peak microamperage was monitored at all times with a calibrated oscilloscope across a 1000Ω resistor (Fig. 38). The sensitive microammeter in series with the preparation served as a null indicator and the variable 500KΩ series resistor allowed rapid adjustment of pulse amplitude without disturbing the balance. Pulse-pair frequency was standardized at 30 Hz and the pulse trains were turned on and off automatically at 30-sec intervals unless otherwise specified.

The Septal-Preoptic-Tuberal System (SPTS) and the Ovulatory LH Surge

The present-day understanding that in female rats gonadotropin secretion has dual controls stems from the report by Hillarp (1949) that large bilateral lesions in the anterior hypothalamic area resulted in constant vaginal cornification and chronic failure of ovulation. This finding, in agreement with similar results described in guinea pigs by Dey et al. (1940) and Dey (1943), has been amply confirmed by various investigators (see review by Kalra and Kalra 1983). On the basis of Hillarp's report Everett and Sawyer (1949, 1950) postulated the existence of apparatus in the rostral hypothalamus having intrinsic circadian rhythmicity and controlling ovulation (see p. 52). The tuberal region and median eminence were judged essential for development of large ovarian follicles and estrogen secretion and also to serve as part of the final common path through which the rostral cyclic apparatus operates. Mapping experiments with the electrochemical method of stimulation gave added weight to this view and indicated a funnel-shaped septal-preoptic-tuberal system that is rostrally diffuse and sharply focussed upon the region of the median eminence (Figs. 36 and 37).

Whereas Critchlow (1958) had found that to induce ovulation electrodes in the medial basal tuber (MBT) must rest very close to the median eminence, we

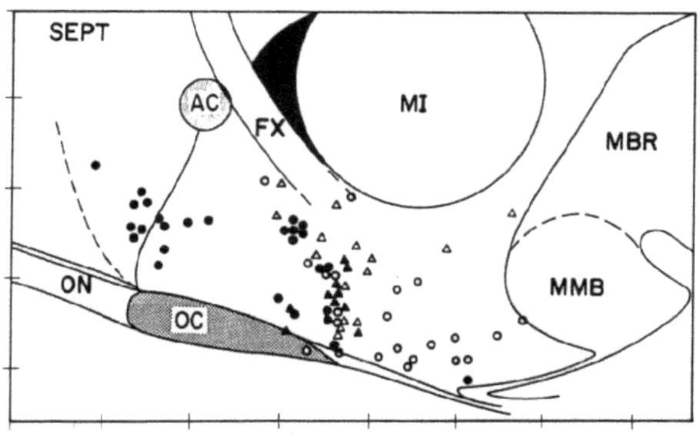

Fig. 36. Locations of *centers* of small irritative lesions in an exploratory study of the rat hypothalamus and preoptic areas. Sagittal diagram. Black symbols: induced ovulation; open symbols: negative; circles: 1 mm from midline; triangles: 1.1 to 2.5 mm lateral. Proestrous 4-day cyclic O-M rats. Parameters: 100 μA, 1 msec anodic pulses at 100 Hz for 60 sec. (Everett et al. 1964)

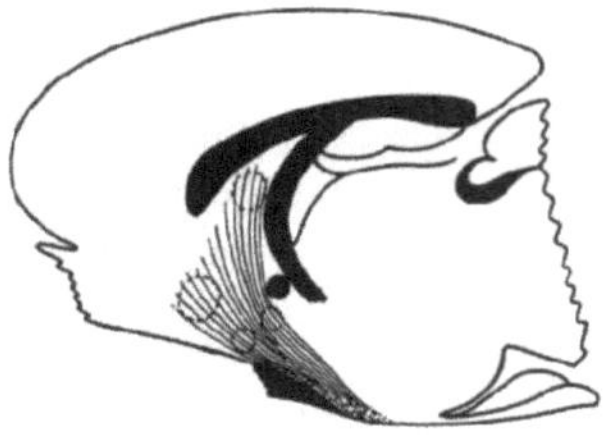

Fig. 37. Diagrammatic parasagittal view of the rat brain showing the postulated septo-preoptic-tuberal system concerned with ovulation. Widely dispersed origins in the septal complex are indicated in the septal complex (left), where relatively large irritative lesions are required compared with the preoptic area (broken lines). (From Everett et al. 1964)

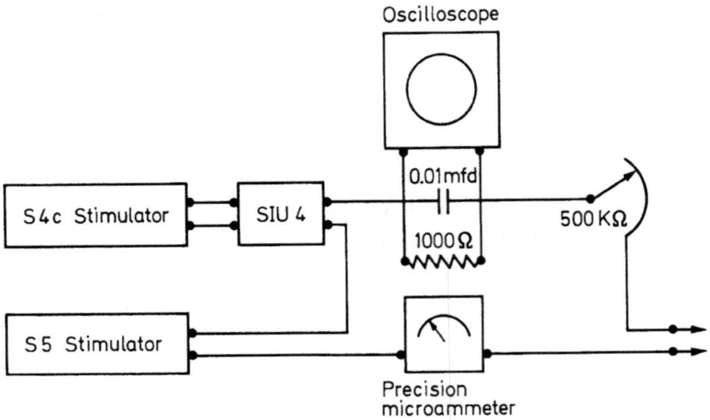

Fig. 38. Wiring diagram of the stimulator circuit. For *electrochemical* stimulation only the S4c/SIU4 was used for anodic monophasic pulses or direct current. For *electrical* stimulation each half pulse from that unit was matched by one from the S5 with opposite polarity

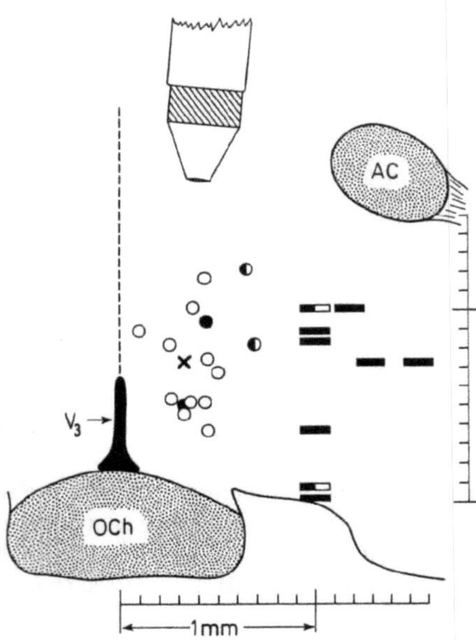

Fig. 39. Diagrammatic coronal section of the preoptic region, showing electrode positions in two groups of CD proestrous rats stimulated with 500 μA pulse pairs for 45 min. Coaxial electrode locations marked by circles (o = ovulation; o = follicle activation; o = no ovulation; x = estimated position, no ovulation. Positions of the right member of the spaced electrode pair are marked by the black bars; black-white: follicle activation, no ovulation; black: negative. A coaxial electrode is drawn to scale. (From Everett et al. 1976)

discovered (Everett et al. 1964) that ECS was almost uniformly effective in the MPOA wherever the electrodes were placed (Fig. 36). In that exploration the EC lesions were made unilaterally with 100 μA *anodic* rectangular 1.0 msec pulses at 100 Hz for 60 sec (600 μcoulombs). The EC lesions observed on the next day were 0.4–0.5 mm in diameter, approximately that of the anterior commissure. In the suprachiasmatic frontal plane effective electrode sites ranged laterally 1.0–2.5 mm well into the lateral hypothalamus (Fig. 40) so as to involve the medial forebrain bundle (MFB). (Much of the communication from the septum and MPOA to the retrochiasmatic hypothalamus comes by way of the MFB [Conrad and Pfaff 1976; Merchenthaler et al. 1980; Szentagothai et al. 1968]). Sites in the lateral POA were not effective.

Several of the very rostral effective sites depicted in Fig. 36 were actually anterior to the MPOA in the diagonal band of Broca. Exploration still farther forward and dorsally into the septal complex with large 1.5–2.0 mm EC lesions (100μA DC, 30 sec) gave uniformly positive results from locations in the medial and lateral septum, the nucleus accumbens and the medial parolfactory area (Fig. 37). Although the SPTS was originally conceived to be an array of possibly continuous axons, that view has required modification in view of the evidence cited earlier that the stimulative action of an EC lesion is exerted on perikarya but not on axons. Nonetheless, the general concept of a rostrally dispersed system appears to be sound.

The rostral territories where small EC lesions act effectively are neither anatomically nor funtionally homogeneous, however. Selective elimination of different rostral regions by destructive lesions produces different effects. Thus, depending on the location and size of lesions that produce persistent estrus, progesterone treatment may or may not induce ovulation (Barraclough et al. 1964; Terasawa et al. 1980). Clemens et al. (1976) produced sequences of pseudopregnancies with large bilateral lesions of the MPOA rostral to the suprachiasmatic nuclei, apparently sparing the small basally located medial preoptic nuclei (MPN). According to Terasawa et al. (1980) restricted bilateral destruction of the MPN caused persistent estrus that could not be overcome by moderate ECS in the bed nucleus of the stria terminalis or in the MPOA. On the other hand, rats made persistent-estrous by destruction of the periventricular part of the MPOA or of the suprachiasmatic region caudal to the MPN, consistently ovulated in response to such stimulation.

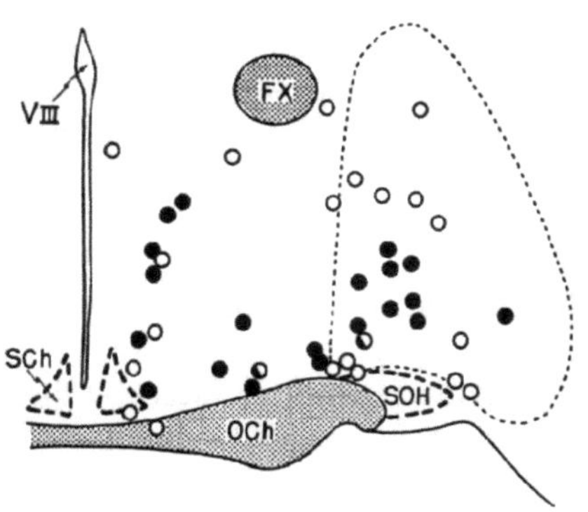

Fig. 40. Diagram of a coronal section of the hypothalamus at the suprachiasmatic plane showing positions of the electrode tips in attempts at electrochemical stimulation of ovulation. Key: FX = fornix; OCh = optic chiasma; SCh = suprachiasmatic nucleus; V_{III} = third ventricle. Broken line circumscribes lateral hypothalamus. Black circles: positive. White circles: negative. Black and white circles: partial effect. (From Everett 1969)

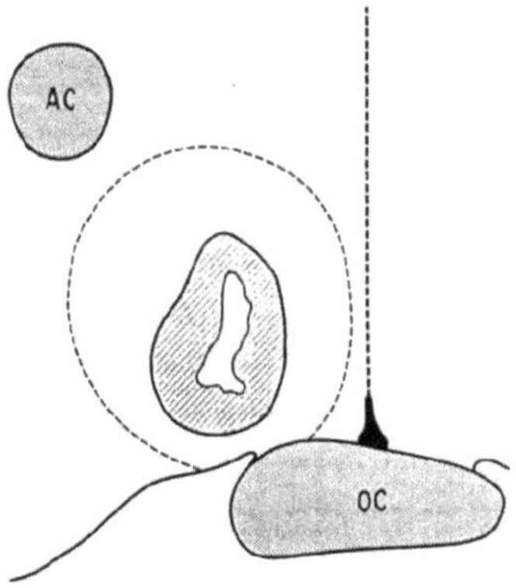

Fig. 41. Camera lucida tracing of a coronal section of the preoptic brain outlining a characteristic electrochemical lesion produced by anodal current of 233 μA for 30 seconds. The central coagulated zone around the empty electrode site is hatched. The outlying zone limited by the thin broken line when observed on the day after electrolysis is pale staining and devoid of neural perikarya. During the first hours after passage of current this zone retains histologically intact neurons and is presumably the site of excitatory action of iron deposited from the electrode. The amount of LH released is correlated with the size of the lesion as well as the amount of electric current. (From Everett and Tyrey 1982b)

Further evidence of wide dispersion of the SPTS in the suprachiasmatic frontal plane became evident in studies by Halász and Gorski (1967) and Tejasen and Everett (1967). The former authors, making use of a bayonet-shaped knife as devised by Halász and Pupp (1965) for stereotaxic surgery of the rat hypothalamus, made a variety of partial or complete deafferentations of the MBT, noting the effects on spontaneous ovulation. Sweeping cuts along the caudal margin of the optic chiasm extending laterally 1.5 mm resulted in chronic blockage of ovulation. Shorter cuts were not effective at that level, but blocked ovulation if placed 1.5 mm behind the chiasm. Tejasen and Everett (1967), working with pentobarbital-blocked proestrous rats, examined whether the ovulatory effect of a standardized unilateral MPOA ECS could be prevented by transverse cuts behind the stimulation site made ipsilaterally or contralaterally prior to the stimulation. Ipsilateral cuts consistently blocked ovulation, provided that they extended to the base of the brain and laterally to include tissue between 0.3 and 1.4 mm from the midline (Fig. 42). Cuts that began 0.4 or 0.5 mm from the midline or that extended only 1.0 mm laterally failed to block. Contralateral cuts also failed.

Non-Specific Stimulation of the Tuber by Destructive Lesions

Within the tuberal region, LH release and ovulation can be induced by tissue damage itself, from either bilateral surgical cuts (Tejasen and Everett 1967) or radiofrequency (RF) lesioning (Everett and Tyrey 1977). Fig. 43 shows the dis-

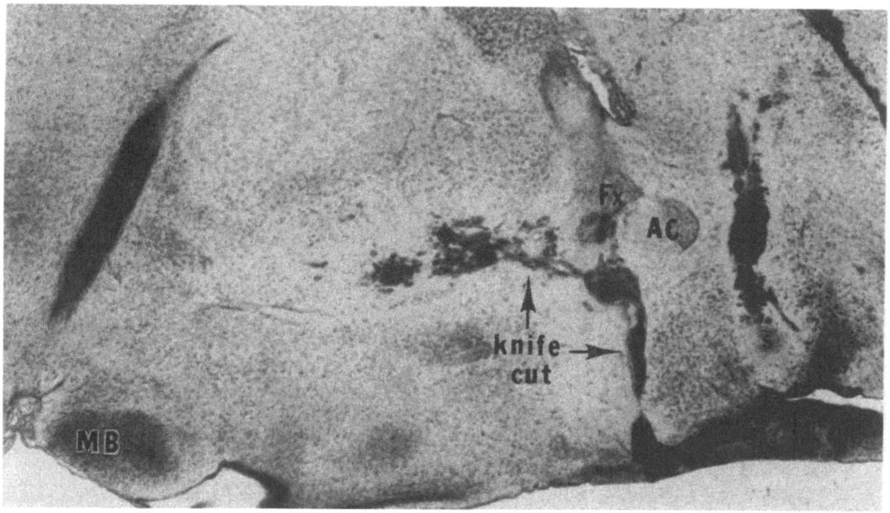

Fig. 42. Characteristic appearance in parasagittal section of the rat brain cut with a Halasz-type knife and electrochemically stimulated in the MPOA rostral to the cut. This was a unilateral cut, ipsilateral to the stimulation site. Ovulation failed to occur. Abbreviations: AC = anterior commissure; Fx = fornix; HP = habenular-interpeduncular tract; MB = mammillary body; S = stimulation site. (From Tejasen and Everett 1967)

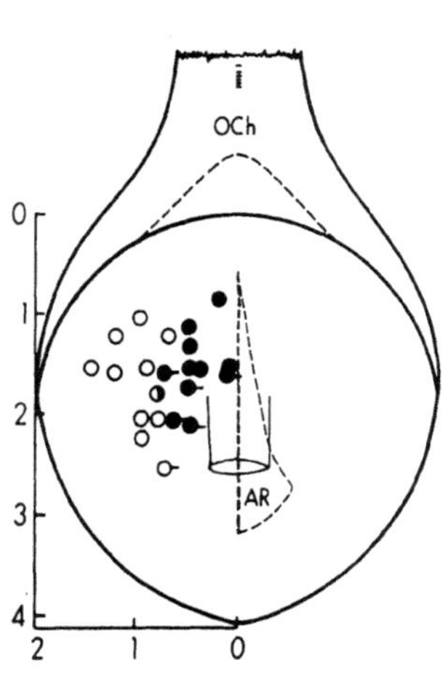

Fig. 43. Locations of the centers of radiofrequency lesions near the floor of the tuberal hypothalamus, representing in black circles those that induced LH release and ovulation and in white those that failed: (○ ●) = unilateral lesions; (○– ●–) = one member of a pair of bilateral lesions nearest the midline (electrode pair spaced 2 mm apart). Scales in millimeters. AR: arcuate nucleus

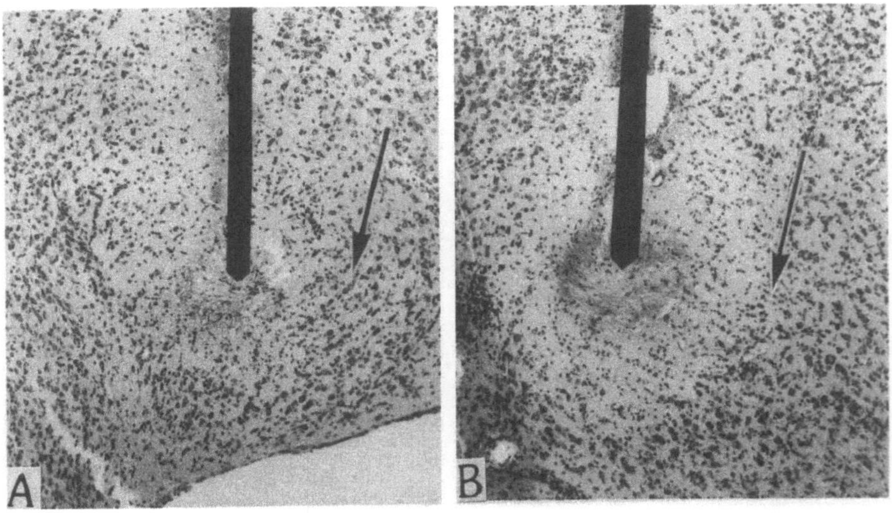

Fig. 44. Comparison of electrochemical lesions of minimal effective size introduced, respectively, during proestrus of the 4-day cycle (A) and diestrus day 3 of the 5-day cycle (B). O-M rats. The pointed black bars follow the electrode tracks to the centers of the lesions. Arrows mark the approximate outer limits of the neuron-free zones. The darkened coagulated areas close to the electrode tips represent heavy deposits of iron demonstrated by the prussian blue reaction. (From Everett 1964)

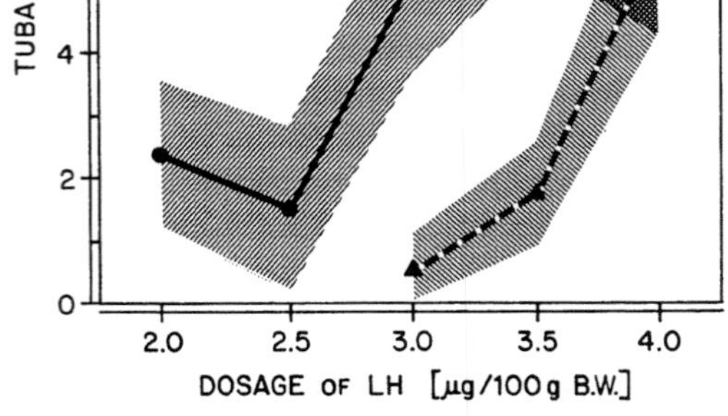

Fig. 45. Relative effectiveness of various dosages of bovine LH (sc) for inducing ovulation in pentobarbital-blocked O-M rats during proestrus of the 4-day cycle and diestrus day 3 of the 5-day cycle (see Table 17)

tribution of RF lesions that produced ovulatory surges of LH. The electrodes for this purpose were the coaxial platinum-iridium units previously described for electrical stimulation (p. 62), only the core serving as the active element. The indifferent element, a brass rod, was inserted in the rectum. The 2 MHz current produced lesions in the basal tuber ~0.5–0.7 mm in diameter. Bilateral lesions centered about ~1.5–2.0 mm caudal to the optic chiasm and 1.5 mm apart produced full ovulation in 5 of 5 rats; another failed to ovulate, but her lesions were small and more caudal than the positive cases (2.5 mm behind the chiasm). Unilateral lesions centered 0.5 mm or less from the midline and 0.8–1.5 mm behind the chiasm resulted in full ovulation in 6 of 6 rats. Serum LH concentrations by radioimmunoassay of blood samples taken by cardiac puncture 60 min and 90 min after lesioning were 300 ± 75 and 483 ± 183 ng/ml, respectively (ref. NIAMDD LH RP_1). The fact that LH concentration continued to rise long after the brief lesioning procedure suggests continuing irritation; alternatively, there may have been initial dumping of a comparatively large amount of LHRH, equivalent to a bolus injection. It should be recalled that McCann et al. (1960, 1964) reported suggestive evidence of LH release resulting from piqûre of the median eminence and other parts of the rat brain.

Because of the non-specific stimulative action of tissue damage in the basal tuber, it seems unwise to attempt comparison of responses to ECS of the MBT and the MPOA. Electrical stimulation affords more satisfactory comparison.

Dose-Response Relationship of Stimulation to LH Release and Ovulation

The size of an EC lesion determines the amount of LH released and its rate. The first evidence came from attempts to induce ovulation in D-3 rats and to compare the effectiveness of different intensities of stimulus between D-3 rats and proestrous rats. The D-3 rats required about twice as much stimulation (4×10^{-4} coulombs anodal D.C.) as proestrous rats (Table 16) (Everett 1964; Everett et al. 1964). The comparative sizes of the minimal effective EC lesions are represented in Fig. 44. A coordinate study in the two categories of animals was made of the ovarian responses to various subcutaneous dosages of bovine LH under pentobarbital blockage (unpublished). As shown in Table 17 and Fig. 45, given doses of LH were significantly more effective in proestrous than in diestrous subjects. This accounts in part for the difference in the requirements for MPOA stimulation. But there are also differences at the hypothalamopituitary level.

The AP is more responsive to given amounts of LHRH injected on proestrus than on D-3 of the 5-day cycle, as measured by serum LH concentration (Martin et al. 1974a). Priming 5-day rats with either EB administered on D-2 or progesterone at noon on D-3 three hours before injection of LHRH, increased the responsiveness on D-3 to the proestrus level (Martin et al. 1974b). Conversely, in 4-day rats when diestrus was artificially prolonged one day by injecting progesterone on D-1, the responsiveness 2 days later was reduced.

The comparative lengths of time needed for release of the ovulatory quota of LH were examined after a standard level of ECS in 5-day proestrous rats and 5-day diestrus-3 rats, repectively (Everett 1964). Under pentobarbital blockade, ECS of the

Table 16. Comparative effectiveness for MPOA electrochemical stimulation with low levels of direct current during late diestrus and proestrus in O-M rats. (Modified from Everett et al. 1964)

μA D.C.	Time (sec)	μcoulombs	Results[a]	
			Diestrus	Proestrus
10	60	600	+ + +	
	50	500	+	
	40	400	+ + + +	
	30	300	00	+
	20	200	00000	+ + + + + + + +
	15	150		+ + 00
	10	100		+ 0000
7.5	10	75		00000

[a] Each symbol represents an individual rat (+ = ovulated, 0 = not ovulated).

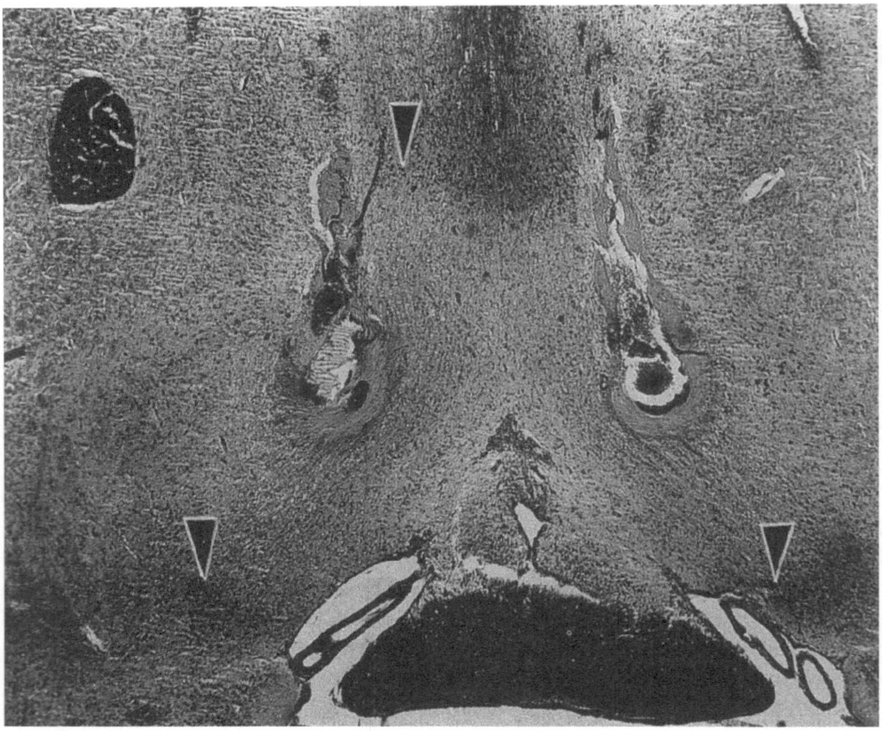

Table 17. Comparative responses of proestrous and late-diestrous rats to bovine LH

LH dose μg/100 g body wt sc	Proestrus 4-day cycle		Diestrus day 3 5-day cycle	
	No. rats	Av. no. ova	No. rats	Av. no. ova
2	8	2.4 1.2[a]	4	0
2.5	6	1.5±1.3	0	0
3	8	5.2±1.5	8	0.5±0.6
3.5	11	6.7±1.2	7	1.7±0.8
4	5	6.4±2.1	10	5.8±1.4

[a] ±S.E.

Table 18. Rate of release of ovulation-inducing hormone as measured by hypophysectomy at different intervals after initial stimulus. (From Everett 1964)

No. of rats	Stage of Cycle	Lesioning current	Interval to hypoX (min)	Effect on ovaries			
				None	Partial	Compl. ovulation	Av. No of ova
5	Proestr.	Pulses[a]	28–32		5		0.2
13	Proestr.	Pulses[a]	40–47		4	9	7.2
10	Diestr.-3	Pulses[a]	38–43	3	7		1.0
4	Diestr.-3	Pulses[a]	60–64		1	3	6.7
7	Proestr.	D.C. dual[b]	29–31		3	4	5.6

[a]60-sec train of 150 μA, 1 msec rectangular monophasic pulses, 100/sec.
[b]Direct current, 100 μA, 30 sec, through separate electrodes to each side.

MPOA was administered unilaterally with a 60-sec train of 150μA, 1 msec, rectangular, monophasic, anodic pulses at 100 Hz (900 μcoulombs). Parapharyngeal hypophysectomy was performed at various times thereafter. Most of the proestrous rats hypophysectomized at 40–47 min ovulated completely by the next morning (Table 18). In the diestrous rats that degree of response appeared only when hypophysectomy was delayed until 60–64 min after the stimulus.

Fig. 46. Coronal section through the preoptic region showing large, bilateral, nearly symmetrical electrochemical lesions produced by 100 μA D.C. for 30 sec separately to each side. Stimulation during proestrus; hypophysectomy at 30 min; 11 tubal oval at necropsy on the following morning. Arrows indicate the approximate outer limits of the neuron-free zones. Optic chiasma below; rostral tip of 3rd ventricle immediately above it. Anterior limbs of the anterior commissure at the upper corners. (From Everett 1964)

Another aspect of that study relates to the fact that atropine treatment or hypophysectomy during the proestrus critical period indicated that in the spontaneous LH surge the release of the ovulatory quota occupies only about 30 min (Everett and Sawyer 1953; Everett 1956). In the present instance, it appeared that after the moderate level of ECS (900 μcoulombs) release of that quota required a longer time. It seemed likely that a larger stimulus might release the necessary amount within 30 min. Consequently, 7 proestrous rats were given bilateral ECS with 100 μA D.C. for 30 sec (3000 μcoulombs). The resulting lesions consumed most of the MPOA (Fig. 46). Hypophysectomy was accomplished at 29–31 min. Of the 7 rats, 4 presented full ovulation next morning and the other 3 showed clear histological evidence of ovarian activation.

Thus, after ECS the amount of LH released within a given time varies with the amount of electricity delivered, with the size of the MPOA lesion and the extent of involvement of the MPOA region. In order to approximate the rate of LH release in the spontaneous proestrous surge, ECS of the MPOA must involve essentially all of the SPTS.

Up to this point, the neural stimulation studies were limited to O-M rats. Subsequent use of CD rats revealed that they required substantially greater levels of MPOA stimulation than O-M rats (Everett et al. 1970). Table 19 displays the respective responses of the two strains as measured by the numbers of tubal ova produced. Whereas full ovulation was uniform among O-M rats stimulated with 10 μA for 20 sec (200 μcoulombs), to approximate that result in CD rats required at least 75 μA for 30 sec (2250 μcoulombs). The strain differences and the dose-response

Table 19. Comparative preoptic thresholds to electrochemical stimulation in 4-day cyclic proestrous rats of two strains. (Modified from Everett et al. 1970, Tables I and II)

Strain	Direct current (μA)	Time (sec)	μcoulombs	Ovarian effect		
				Full[a]	Partial[b]	None
O-M	7.5	10	75	0	1	1
	10	10	100	1	0	4
	10	15	150	1	1	3
	10	20	200	6	0	0
	15	10	150	3	1	0
CD	10	20	200	0	1	1
	10	30	300	1	2	3
	10	60	600	4	3	1
	20	30	600	2	4	3
	20	60	1200	4	0	3
	50	30	1500	6	3	1
	75	30	2250	7	2	1
	100	30	3000	5	1	0

[a] Full ovulation: 7 or more tubal ova.
[b] < 7 tubal ova or only histologically detectable lutein changes.

Table 20. Comparative relationships in O-M versus CD rats of the extent of preoptic electrochemical stimulation, amounts of circulation LH at 90 min. and the resulting ovulations. (From Everett et al., 1973)

Strain of rat	Stage of cycle[a]	Stimulation μA×sec	Stimulation μCoulombs	Serum LH ng/ml mean ± SE	Ovulation-Luteinization[b] Full	Partial[c]	Absent
O-M	P	100×30	3000	832 ± 154	5	0	0
		10×40	400	163 ± 23	5	0	0
	D-3	100×30	3000	204 ± 23	4	1	0
		10×40	400	77 ± 10	2	0	3
CD	P	100×30	3000	409 ± 33	9	0	0
		20×40	800	129 ± 13	4	2	3
	D-3	100×30	3000	172 ± 28	5	1	3
		20×40	800	88 ± 8	3	0	5

[a] P = proestrus in the 4-day cycle, and D-3 = diestrus day 3 in the 5-day cycle.
[b] Numerals indicate numbers of rats in respective categories.
[c] Partial effects range from limited prelutein changes in unruptured follicles to the presence of fewer than 8 tubal ova.

relationships were again noted by Everett et al. (1973) with respect to both the amounts of LH released by differing levels of ECS and the resulting ovarian responses (Table 20).

Electrical Stimulation of MPOA versus Basal Tuber

The first experiments with electrical stimulation made use of the dual platinum electrodes spaced 2 mm across the midline, supposedly to spread the stimulus through much of the MPOA (Everett 1965). In control rats the electrode assembly was moved laterally to span the LPOA (Figs. 47 and 48). The subjects were 4-day cyclic O-M rats under pentobarbital blockade during the proestrus critical period. Once the electrodes were in place, biphasic pairs of 1.0 msec rectangular pulses 1000 μA peak-to-peak were passed for various lengths of time as shown in Table 21. In the MPOA group, with increase of stimulation time there was a progressive increase in numbers of rats ovulating and numbers of ova produced, until full ovulation was reached in rats stimulated for 45 and 60 min. Stimulation in the LPOA for 60 min was rarely effective. When peak-to-peak current was lowered to 500 μA for 45 min, the proportion of rats induced to ovulate dropped to 57% (4/7) and only one showed full ovulation. Further reduction to 200 μA for 45 min gave entirely negative results.

The next step was to compare effects of electrical stimulation of the MBT with findings in the MPOA, using similar parameters of current strength and time (Quinn and Everett unpublished). With the highest current (1000μA) results were much as before: stimulation for 15 min ovulated 2/5 rats and stimulation for 45 min ovulated 8/8 rats. However, at lower intensities of current the MBT site for stimulation

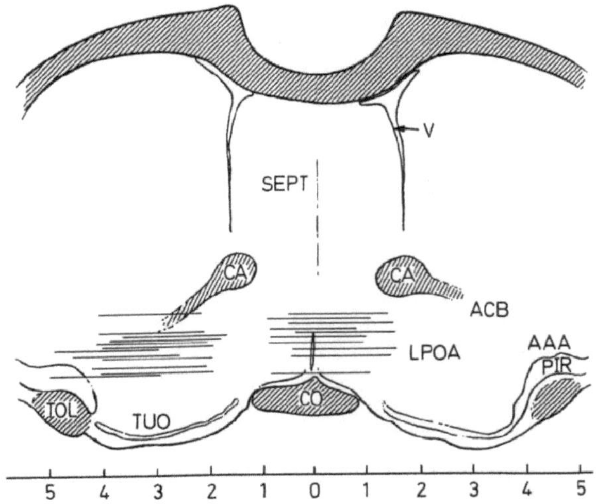

Fig. 47. Diagrammatic coronal section of the rat brain just rostral to the anterior commissure, showing by horizontal lines the span of paired electrodes in various transmedian (positive) and lateral (chiefly negative) electrical stimulation experiments. Key: AAA = anterior amygdaloid area, ACB = nucleus accumbens, CA = anterior commissure, CO = optic chiasma, LPOA = lateral preoptic area, PIR = piriform cortex, SEPT = septum, TOL = lateral olfactory tract, TUO = olfactory tubercle, V = lateral ventricle. Proestrous O-M rats. (From Everett 1965)

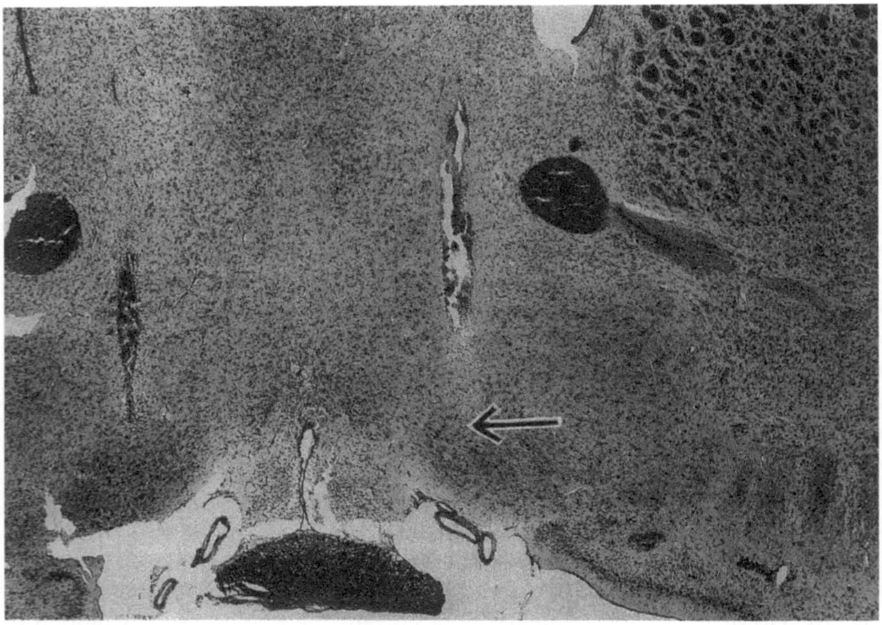

Table 21. Induction of ovulation by preoptic electrical stimulation. Effects of duration of stimulus and electrode position. O-M rats. (Modified from Everett 1965)

Electrode location	Stimulus[b] duration (min)	No. of rats	Numbers of tubal ova				
			None	1–3	7–9	10–12	Mean ± SE[c]
Medial POA[a]	0	5	4	1			0.2 ± 0.19
	10	5	5				
	15	8	5	1	2		2.4 ± 1.4
	20	8	3	2	3		3.5 ± 1.5
	30	8	2		2	4	7.5 ± 1.9
	60	7			1	6	10.7 ± 0.5
	70	1				1	
Lateral POA							
>1.5 mm from midline[d]	60	14	11	2		1	0.9 ± 0.74
<1.5 mm from midline[d]	60	2	1			1	

[a] POA = preoptic area. Electrodes spanning 2 mm across midline.
[b] Biphasic pulse pairs, 30/sec, each half pulse 1 msec, 1 mamp peak-to-peak, 30-sec trains at 30-sec intervals.
[c] Standard error of the mean.
[d] Position of more medial electrode.

showed an advantage over MPOA. In particular, 500 μA pulses for 45 min ovulated 14/14 rats, almost all having full sets of tubal ova. The apparent advantage of MBT sites became certain when the study was extended to the CD strain and the induced LH surges were measured by radioimmunoassay.

Everett et al. (1976) compared the effectiveness of MPOA and tuberal electrical stimulation as measured by ovulation frequencies in both O-M and CD proestrous rats. The investigation further examined the relative efficiencies of an electrode pair spaced 2 mm across the midline and a coaxial electrode placed unilaterally near the third ventricle. Electrode sites in the MPOA are represented in Fig. 39. In the MBT they were close to the brain floor within about 1 mm behind the caudal margin of the optic chiasma. Biphasic pulse trains at various microamperages were delivered for standard durations of 45 min. Tuberal stimulation (with the spaced electrodes) was distinctly more effective for ovulation than MPOA stimulation in both O-M and CD rats (Table 22). The advantage of the tuberal site was especially marked in CD rats that received 500 μA pulses, all 20 rats ovulating, by contrast with uniform failure in the 9 rats stimulated in the MPOA. In comparison of the two strains, O-M rats gave somewhat better responses to given levels of current in the MPOA, whereas results

◄

Fig. 48. Coronal section of brain showing electrode positions in transmedian electrical stimulation of the medial preoptic areas. Tip of the right-hand electrode reached the depth indicated by the arrow in adjacent sections. Stimulation for 30 min; 11 tubal ova present next morning. Proestrous O-M rats. (From Everett 1965)

Table 22. Comparative efficiency of electrical stimulation of the mPOA or tuber in two strains of rats. Biphasic pulse pairs, varying microamperage for 45 min. Platinum electrodes spaced 2 mm across the midline. (From Everett et al. 1976)

		O-M rats		CD rats	
Region stimulated	Stimulus current[a]	Rats ovulated	Ova per ovulation	Rats ovulated	Ova per ovulation
mPOA	1,000 μA	6/6	10.3 ±0.6[b]	7/10	8.6 ±2.3
	500 μA	4/7	4.8 ±1.9	0/9	–
	200 μA	0/7	–	–	–
Tuber	1,000 μA	8/8	10.9 ±0.4	–	–
	500 μA	14/14	8.9 ±0.7	20/20	13.0 ±0.6
	350 μA	7/8	7.4 ±1.6	6/6	11.7 ±2.0
	200 μA	1/5	12[a]	5/10	8.6 ±3.1

[a] Peak-to-peak, pulse pair.
[b] Mean ±SE.
[c] Two others had histologic evidence of partial follicular activation.

from tuberal stimulation were essentially alike in O-M and CD rats at all intensities. The relative effectiveness of the spaced and coaxial electrodes was examined in CD rats (Tables 23–25). In the preoptic sites (Fig. 39, Table 23) the more medially placed coaxial electrode was clearly superior. In the tuberal sites both types of electrode gave comparable results in terms of ovulation (Tables 23 and 24), but the coaxial electrode produced significantly higher levels of serum LH (Table 25).

Table 23. Comparative efficiencies of spaced versus coaxial electrodes in electrical stimulation of the medial basal tuber in 4-day and 5-day proestrous CD rats. (From Everett et al. 1976)

Electrode type	Cycle length	Stimulus current[a]	Rats ovulating	Ova per ovulating
Spaced	4-day	500 μA	20/20[c]	13.0 ±0.6[b]
		200 μA	5/10[d]	8.6 ±3.1
	5-day	500 μA	12/16[c]	8.0 ±1.4
Coaxial	4-day	500 μA	10/10[c]	12.9 ±0.6
		200 μA	7/10[d]	8.4 ±2.2
	5-day	500 μA	8/10[c]	10.9 ±1.0
		350 μA	5/7[d]	11.8 ±0.9
		200 μA	5/9[d]	7.2 ±3.2

[a] Peak-to-peak. Pulse pairs, 45 min.
[b] Mean ±SE.
[c,d] Ratios bearing the same letter are not statistically different.

Table 24. Comparative efficiencies of spaced and coaxial electrodes (pulse pairs, varying microamperage, 45 min) in the preoptic and tuberal regions of 4-day cyclic CD rats. (From Everett et al. 1976)

Region	Stimulus current[a]	Spaced electrode: Rats ovulated	Spaced electrode: Ova per ovulation	Coaxial electrode: Rats ovulated	Coaxial electrode: Ova per ovulation
mPOA	1,000 μA	7/10	8.6±2.3[b]	–	–
	500 μA	0/9	–	11/16	8.2±1.6
	200μA	–	–	4/15[c]	12.8±0.8
Tuber	500 μA	20/20	13.0±0.6	10/10	12.9±0.6
	200 μA	5/10	8.6±3.1	7/10	8.4±2.2.

[a] Peak-to-peak, pulse pair [b] Mean ±SE.
[c] Includes 7 cases in which the pulse frequency was 200 Hz, 2 rats ovulating.

Table 25. Serum LH concentrations 90 minutes after the beginning of tuberal stimulation (500 μA pulses, 45 min), comparing effects of the spaced electrode pair and the coaxial electrode in 4-day and 5-day cyclic CD rats. (From Everett et al. 1976)

Cycle length	Spaced electrodes: N	Spaced electrodes: LH(ng/ml)	Coaxial electrode: N	Coaxial electrode: LH(ng/ml)	Ratio: coaxial/spaced	P
4-day	9	232±46[a]	11	395±83	1.70	<0.05
5-day	13	236±35	7	323±50	1.37	>0.5
4-day + 5-day	22	234±27	18	367±53	1.57	<0.025

Differences between 4-day and 5-day cyclic rats are not significant.
[a] Mean ±SE.

LH Requirements for Ovulation

Gosden et al. (1976) examined the relationship between the level of circulating LH and the length of time it was sustained. The subjects were 4-day cyclic CD rats in proestrus under pentobarbital during the critical period. They received electrical stimulation with the coaxial electrode in either the MPOA (670 μA pulse pairs × 45 min) or the arcuate nucleus-median eminence (ARC/ME) (500 μA pulse pairs × 45 min). Blood samples were obtained through intra-atrial cannulae before and at 45-min intervals after stimulation began. Serum LH rose to moderate levels by 45 min, fell somewhat by 90 min and reached base line at about 180 min (Fig. 49). As before, the LH levels after ARC stimulation averaged higher than after MPOA stimulation. There was no rise after sham stimulations in which the electrode was lowered into position and simply left there for 45 min. There was no clear relationship between the peak levels of LH in rats ovulating and the numbers of ova produced. Full ovulation occurred in rats having maximal LH concentrations as low as 187 ng/ml serum. In most cases LH continued to be secreted after the 45-min stimulus ended. In some cases the 90-min concentrations were equal to or higher than 45 min. The electrode locations were similar to those described by Everett et al. (1976).

In a second group of CD proestrous rats, transauricular hypophysectomy was performed at various times after the start of MPOA or ARC stimulation (Table 26, Fig. 50). Whereas sham hypophysectomy at 45 min failed to interfere with ovulation, removal of the pituitary at that interval prevented ovulation in all cases. Only histologic examination disclosed a sub-ovulatory degree of ovarian activation. By contrast, when hypophysectomy was delayed until 65 min nearly all rats ovulated

Table 26. Effect of time of hypophysectomy on ovulation in pentobarbital-blocked proestrous CD rats receiving unilateral electrical stimulation of the hypothalamus. (From Gosden et al. 1976)

Site of stimulation	Time of hypophysectomy after onset of stimulation	No. of animals		Ova per ovulation
		Total	Ovulating	
	(min)			
Arcuate nucleus	45	6	0[a]	
	55	6	3	4.3 ± 2.8[c]
	65	7	7[a]	10.3 ± 1.3
	75	2	1	13
	45 (sham hypox)	3	3	13.0 ± 1.0
Preoptic area	45	7	0[b]	
	65	7	6[b]	9.0 ± 1.8
	45 (sham hypox)	3	3	12.7 ± 0.7

[a,b] Values having the same superscript are significantly different ($P < 0.01$).
[c] Mean ± SE

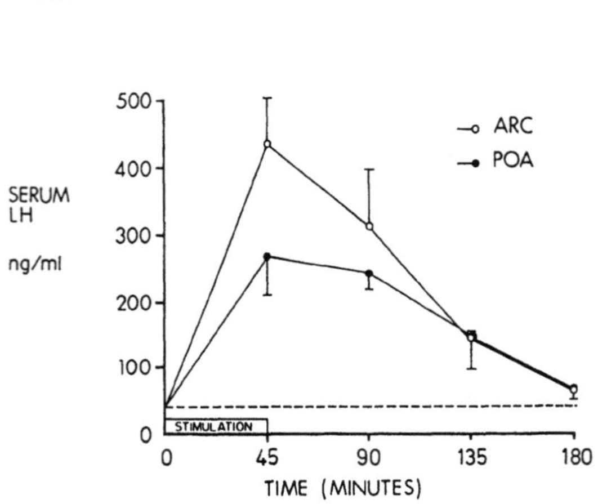

Fig. 49. Serum LH changes with time in proestrous CD rats receiving unilateral electrical stimulation to the arcuate nucleus (6 rats) or the medial preoptic area (7 rats) during the critical period. The SE is represented by the vertical lines above or below the means. LH levels in sham experiments were below the broken line. (From Gosden et al. 1976)

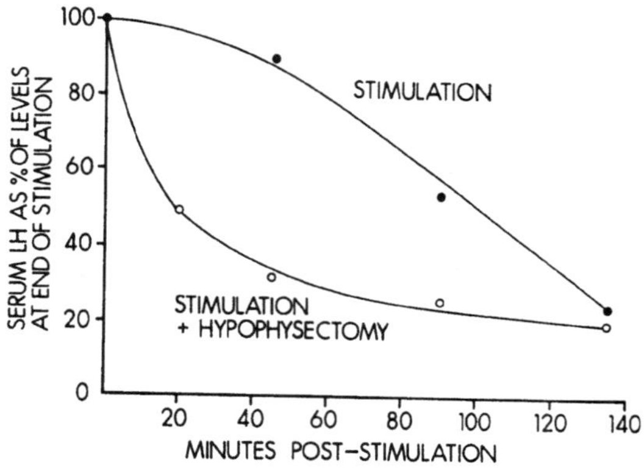

Fig. 50. Serum LH changes in intact rats and in rats hypophysectomized immediately after 45-min MPOA electrical stimulation. LH values expressed as percentage of the mean concentrations at 45 min. (From Gosden et al. 1976)

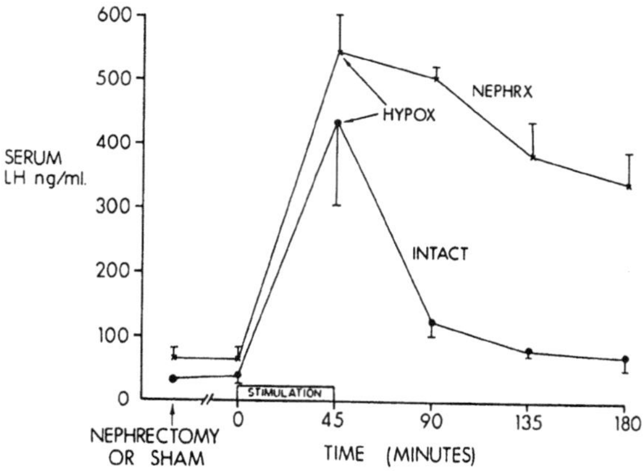

Fig. 51. Serum LH changes in nephrectomized (-x-) or sham-nephrectomized (-o-) proestrous rats hypophysectomized 45 min after onset of MPOA electrical stimulation. Means ± SE. (From Gosden et al. 1976)

and the average number of ova discharged was in the normal range. Hypophysectomy at the intermediate interval of 55 min resulted in ovulation in 3 of the 6 rats and a below-normal average number of ova per ovulation. Thus, with the degrees of stimulation used, LH levels reached at 45 min were inadequate for ovulation and a longer period was essential for the pituitary to release the minimal *ovulation quota*. In unpublished experiments of the same study, increasing the intensity of ARC/ME stimulation (1000 μA pulse pairs × 45 min) produced ovulation of small numbers of ova in 3 of 6 rats hypophysectomized at 45 min.

The importance of sustained action of LH for ovulation was further shown by nephrectomy in rats that were subsequently subjected to MPOA stimulation (670 μA pulse pairs × 45 min) and hypophysectomy at 45 min or 65 min (Fig. 51). The half-life of serum LH in sham nephrectomized rats after hypophysectomy at 45 min was estimated as ~32 min. None of these 6 rats ovulated. In the nephrectomized rats the estimated half-life of LH increased about seven-fold to ~210 min and resulted in ovulation in 6/7 rats. Thus the occurrence of ovulation can be determined not only by a certain minimal concentration of circulating LH, but also by the length of time that this concentration is maintained.

The direct quantitative relationship between the degree of MPOA or tuberal stimulation and the amount of LH released in a given time span is now well documented (Everett 1964; Everett et al. 1973; Turgeon and Barraclough 1973; Velasco and Rothchild 1973; and the present study). That relationship has best been shown with electrochemical stimulation of the MPOA; we have seen that a rate of release approximating that occurring in the early part of normal surge is realized only when most of the diffuse SPTS at the MPOA level is activated electrochemically. The next section shows that electrical stimulation of either the MPOA or the ARC/ME can approximate the normal surge, provided that stimulation continues for approximately 2 hours.

The Comparative Increments of Circulating LH with Increasing Time of Electrical Stimulation of the Medial Preoptic Area or the Arcuate Nucleus-Median Eminence

In the preceding studies, when the effectiveness of electrical stimulation of the MPOA and the tuberal region was compared, it was evident that while the MPOA threshold was higher than that of the ARC/ME, the two regions appeared to act in concert. This proposal was examined in detail by Everett and Tyrey (1981), with respect to both the ovarian responses and the amounts of LH in circulation after stimulation of the two regions for progressively longer times up to 2 hours. All subjects were young adults of the CD strain having vaginal smear records showing at least two regular 4-day cycles immediately preceding the current cycle. On the afternoon of proestrus each received pentobarbital sodium (35 mg/kg body wt.ip) shortly before 1400h. This produced complete blockage of ovulation in 23 of 25 control rats; one exceptional animal showed histologic evidence of partial ovarian activation and the other produced a single tubal ovum without activation of the other Graafian follicles. For stimulation, a coaxial platinum electrode was placed stereotaxically near the midline in either the MPOA or the ARC/ME. The usual matched biphasic pairs of 1 msec pulses at 30 Hz, on and off at 30-sec intervals were delivered for periods lasting 20, 30, 45, 60, 85 or 115 min. Peak-to-peak current was 750 μA in the MPOA and 500 μA in the ARC/ME.

Results were measured by the serum LH concentrations at the end of the various periods of stimulation and 30 min later; blood samples were taken, respectively, from the external jugular vein and by cardiac puncture (Table 27). Additional information was furnished by the numbers of tubal ova present next morning or by the ovarian histology (Table 28). The results of the 20- and 30-min stimulations, not statistically different, were combined. The curvilinear patterns of serum LH concentration in Fig. 52 resemble that characteristically seen in the rising phase of the normal proestrus surges. The amounts of the hormone released during and following stimulation for 45–60 min correspond to the minimal ovulation quota, nearly all rats stimulated for such periods having ovulated completely. It should be recalled that stimulation for 45 min was adequate for ovulation (Gosden et al. 1976), provided that the pituitary remained in place for an additional 20 min. In the present study, the levels reached after stimulation for 20–30 min were higher in 5 rats that fully

Table 27. Comparative serum LH concentrations after MPOA or ARC/ME stimulation for differing lenghts of time. Proestrous CD rats. (From Everett and Tyrey 1981).

Electrode location	Stimulus duration (min)	No. of rats bled	Serum LH, ng/ml ±SEM, after first pulse				
			25–35 min	60 min	90 min	120 min	150 min
MPOA	20–30	13	143 ± 10	145 ± 14	–	–	–
	45	7	–	442 ± 96	292 ± 56	–	–
	60	7	–	498 ± 88	422 ± 62	–	–
	85	9	–	–	1243 ± 220	762 ± 190	–
	115	7	–	–	–	2167 ± 404	1306 ± 149
ARC/ME	20–30	15	275 ± 62	179 ± 45	–	–	–
	45	12	–	–	421 ± 83	–	–
		10		670 ± 114	457 ± 58	–	–
	85	13	–	–	1859 ± 234	1054 ± 135	–
	115	8	–	–	–	2621 ± 301	1503 ± 227

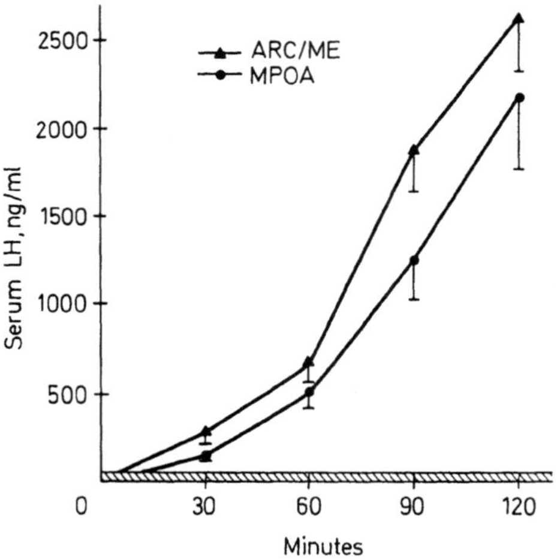

Fig. 52. Serum LH concentrations at the initial bleeding times after electrical stimulation of the ARC/ME or MPOA for progressively longer periods. Each point represents the corresponding mean, and the vertical line shows the SE. Results are in terms of the LH reference preparation RP-1. (From Everett and Tyrey 1981)

Table 28. Comparative ovarian responses after MPOA or ARC/ME stimulation for differing lengths of time. Proestrous CD rats. See Table 27. (From Everett and Tyrey 1981)

Electrode location	Stimulus duration (min)	No. of rats	Ovarian effect		
			None	Partial[a]	Full ovulation
MPOA	20–30	36	13	19	4
	45–60	14	0	5	9
	85	9	0	1	8
	115	7	0	1	6
ARC/ME	20–30	21	7	8	6
	45	22	0	1	21
	85	13	0	4	9
	115	6	0	2	4

[a] Fewer than 9 tubal ova, or histological evidence of luteinization or prelutein maturation.

ovulated than in 7 rats that presented only partial ovarian activation (460 ± 155 ng/ml *vs* 216 ± 46 ng/ml). The much greater amounts of LH released during and after the second hour of stimulation resemble the great excess commonly appearing in spontaneous proestrus surges. The peak concentrations of LH at the end of the 2-hour stimulations (2167 ± 404 ng/ml and 2621 ± 301 ng/ml) are not unlike the highest normal levels. It is of some interest that the LH concentration at the end of 85-min MPOA stimulation (1243 ± 220 ng/ml) is almost identical to that obtained 90 min after massive unilateral ECS of the MPOA (230 μA direct current × 30 sec; 1270 ± 173 ng LH/ml; n = 9).

The incremental curves in Fig. 52 are essentially parallel. Regression analysis of the two sets of data for LH concentration and time of stimulation indicate that the relationship can be represented by straight-line functions with no significant deviation from linearity. The slopes of the regression lines (21.8 ± 5.4 and 26.9 ± 5.2 for the MPOA and ARC/ME data, respectively) are not different statistically ($P > 0.1$).

In spite of the fact that the peak-to-peak current for MPOA stimulation was deliberately set higher than for ARC/ME stimulation, the latter site was consistently more effective in terms of the amount of LH released in a given time. Analysis of variance disclosed a highly significant advantage for the ARC/ME site ($P < 0.01$). The ovulation data were somewhat complicated by the unexpected finding that 8 rats among the 35 stimulated for 85 or 115 min failed to ovulate completely in spite of high levels of serum LH (Tables 27 and 28). Especially puzzling was the extreme case of the rat lacking any sign of ovarian activation after MPOA stimulation for 85 min. Her LH concentrations at 90 and 120 min (2456 and 1739 ng/ml, resp.) were among the highest recorded.

The curvilinear patterns of LH concentration in Fig. 52 agree well with the fact that LHRH is self-priming for its own secretion (Aiyer et al. 1974; Fink et al. 1976; Pickering et al. 1979). After Aiyer et al. demonstrated that when two iv injections of LHRH were given 60 min apart, the second treatment caused a 6-fold increase in the amount of LH released, Fink and associates observed a great increase in the LH response to the second of two 15-min periods of electrical stimulation of the MPOA separated by 45 min. While our investigation was in progress Sherwood et al. (1980) reported the effects of electrical stimulation with spaced electrodes across the MPOA in proestrous rats under urethane anesthesia. When stimulation continued for 6 hours, LH concentration rose rapidly after the first 30 min, reached its peak at 90 min, remained high for another 30 min, and then slowly declined. Our results differ somewhat by showing a continued marked rise of LH from 90 to 120 min, whereas Sherwood et al. showed no significant change. The reason for the difference is obscure.

The striking parallel of the patterns of LH release resulting from MPOA and ARC/ME stimulations suggests that the entire SPTS serves as a functional unit. It is as if stimulation in either area directly activates the LHRH pathway. But even today (1988) that remains speculative. It is significant, however, that the entire system remains excitable under the influence of several drugs and under several physiological conditions that inhibit the spontaneous surge.

Similarity of LH Surges Induced by MPOA Stimulation in Rats Blocked with Various Drugs

In this investigation (Everett and Tyrey 1983) adult 4-day cyclic CD rats in proestrus were subjected to either electrochemical or electrical stimulation of the MPOA after ovulation-blocking treatment with pentobarbital (PTBL), morphine (MPHN), chlorpromazine (CPZ), or atropine (ATR).

One or another of these drugs was injected shortly before 1400h. The dosage for PTBL was 3.5 mg/100 g BW ip. MPHN sulphate was prepared on the day of use (50 ml/mg in 0.9% NaCl) and the standard ip dose was 5 mg/100 g BW. CPZ (Thorazine, 25 mg/ml, Smith, Kline & French, Philadelphia, PA) was diluted on the day of use by 0.9% NaCl to a concentration of 10 mg/ml. The standard dose was 1 mg/100 g BW ip. The rats treated with MPHN or CPZ sometimes required occasional brief ether supplement during stereotaxic surgery, stimulation and blood sampling. In the case of ATR, rats to be electrically stimulated also received PTBL (3.5 mg/100 g BW) for anesthesia during the long periods of stimulation. Since the combination of the two drugs proved unduly stressful for overnight experiments, these animals were killed immediately after the final blood samples were taken. For ECS, the ATR-blocked rats were briefly anesthetized with ether during the surgery, stimulation and blood sampling.

The electrode for ECS was the core of a coaxial stainless steel unit as anode; a brass rod in the rectum served as cathode. The standard stimulus was 233 μA $\times$ 30 sec (~7000 μcoulombs). For ELS the electrode was the usual coaxial platinum-iridium unit and the stimulus consisted of matched biphasic pulse pairs with peak-to-peak current of 750 μA at 30 Hz. Sham controls were prepared stereotaxically in the usual way as if for electrode insertion; in 4/9 CPZ controls, the electrode was inserted and left in place for 60 min.

Blood samples for RIA were usually drawn from the exposed right external jugular vein; some final samples were taken by cardiac puncture. Excepting the ATR + PBTL group and a few rats that died overnight, biological effects were assessed at laparotomy on the morning after stimulation.

Electrochemical stimulation

Table 29 summarizes the effects of unilateral ECS. Of particular interest is the similarity of the average serum LH concentrations 90 min after the brief electrolysis. In first trials with MPHN in sham experiments, 3/9 rats ovulated in spite of the drug; consequently, this series of sham controls was expanded to 18 rats, blood samples being taken from the last 9 rats shown in Table 29. All 15 rats that were stimulated under MPHN ovulated and all but one presented full sets of tubal ova. The exceptional rat had only 4 ova, presumably the result of a minute lesion near the very base of the brain. Although blood samples were taken from 6 of the fully ovulating animals, it was necessary to exclude 3 of them from Table 29 for various reasons: One rat had only a small basal lesion and her 90-min LH concentration was far below the others (410 ng/ml serum). In another, the lesion was centered in the nucleus of the

Table 29. Release of LH and ovulation induced by electrochemical stimulation of the MPOA in rats blocked with MPHN, CPZ, ATR, or PTBL. (From Everett and Tyrey 1982b)

Blocking agent	No. of rats	Serum LH (ng/ml)[a]			No. of tubal ova
		0 time	60 min	90 min	
PTBL[b]	9	< 50		1270 ± 73	10–13
MPHN	3		1025 ± 271	1486 ± 371	11–15
MPHN (sham)	9	95 ± 16		76 ± 6	0[c]
CPZ	4	53 ± 7	633 ± 262	1280 ± 373	13–16
CPZ (sham)	9	62 ± 5	57 ± 5	55 ± 5	0
ATR	10		975 ± 78	1451 ± 107	2–15[d]

[a] Expressed as the mean ± SEM.
[b] Data from Everett and Tyrey (1981).
[c] One rat died overnight; among 9 other sham controls that were not bled, 3 ovulated (13–15 ova). See text.
[d] Eight rats survived overnight.

stria terminalis (1390 ng LH/ml). The third exceptional rat had a lesion extending into the opposite MPOA (1162 ng LH/ml). Of the 3 rats that were included, only one required ether supplements; her 90-min LH concentration was intermediate between the other two.

Among the CPZ-blocked rats, none required supplemental anesthesia for ECS. The 90-min LH values ranged from 930–2275 ng/ml. Full sets of tubal ova were present the next morning.

All of the ATR-blocked rats received the ether supplement for the stereotaxic surgery and stimulation, as well as briefly for each bleeding. Of the 10 rats, one was killed and one died overnight. Among the 8 survivors, 7 ovulated 9–15 ova; the unusual animal produced only 2 ova, but all unruptured follicles were luteinizing. Her 90-min LH concentration was the lowest of the group (994 ng/ml), but a level usually sufficient for full ovulation.

Fig. 53 locates the electrode tips in the MPHN, CPZ and ATR experiments. The lesions typically occupied most of the MPOA on the one side.

Electrical stimulation: Periodic 5-min pulse trains

Because MPHN-blocked rats would tolerate 750 μA pulse trains for only a few minutes without struggling, the usual parameters were changed. Dyer and Mayes (1978) had shown that for inducing ovulation in PTBL-blocked rats, a few 5-min bursts of pulses during 45 min were as effective as continuous stimulation. Confirming that, we also determined that these two methods of stimulation produced equivalent amounts of LH (Fig. 54). During each 5-min stimulation the pulses were automatically turned on and off at 15-sec intervals instead of the 30-sec intervals used in previous studies. The results show that four 5-min trains of pulse pairs in 60 min were as effective as 60 min of continuous trains, both as to the amounts

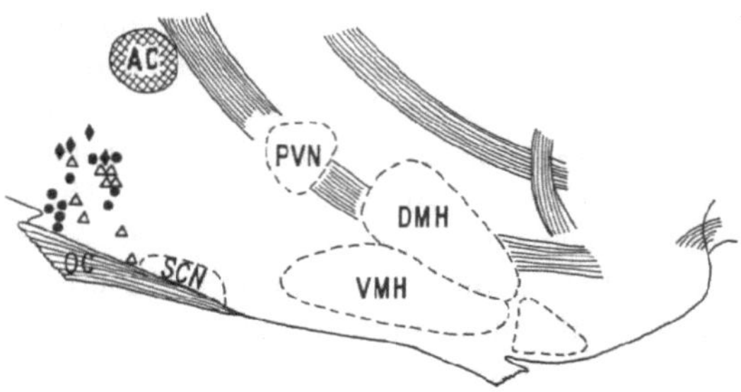

Fig. 53. Approximate electrode sites for electrochemical stimulation in rats blocked with morphine (●), chlorpromazine (♦), or atropine (Δ), projected on a parasagittal diagram of the hypothalamus. Five rats from the MPHN group are omitted here, including three that were mentioned in the text, another with an unusually small basal lesion, and one whose brain was not examined histologically. AC, anterior commissure; DMH, dorsomedial nucleus; OC, optic chiasm; PVN, paraventricular nucleus; SCN, suprachiasmatic nucleus; VMH, ventromedial nucleus. (From Everett and Tyrey 1982b)

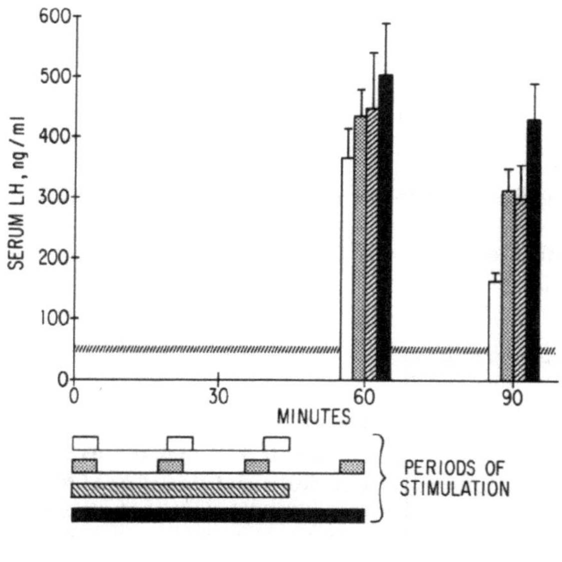

Fig. 54. Serum LH concentrations resulting from MPOA electrical stimulation in proestrous rats blocked with pentobarbital, comparing effects of 5-min bursts of pulse trains with the effects of continuous stimulation for the same overall durations. Each vertical bar, shaded to correspond with one of the stimulation procedures, represents the mean with its standard error above. Basal levels of LH fell below the hatched line. (From Everett and Tyrey 1982b)

of LH released and the numbers of rats tpulating. Stimulation with three 5-min trains in 45 min seemed slightly less effective for ovulation than continuous 45-min pulse trains; only 2 of 7 rats ovulated fully, a result possibly correlated with a precipitous (55%) decline of their LH levels after the 60-min sampling. Compared with the moderate declines in the other groups 3 and 4, that abrupt loss is highly significant ($P < 0.005$). The parameters used in Group 2 were standardized for comparisons among the rats blocked with MPHN, CPZ and ATR + PTBL.

The results of these comparisons are displayed in Table 30 and Figs. 55 and 56. The mean serum LH concentrations were similar among the PTBL, MPHN, and CPZ groups. The rats blocked with ATR + PTBL had somewhat lower values at both 60 and 90 min, but the only difference that was significant statistically was that between the 90-min values of the MPHN and the ATR + PTBL groups ($P < 0.05$). Three of the 7 rats in the ATR + PTBL series are responsible for the low average, for they failed to show LH values higher than 181 ng/ml and would probably not have ovulated had they survived overnight. In the other 4 rats, the 60-min and 90-min averages (421 ± 96 ng/ml and 380 ng/ml, respectively) compared favorably with those obtained in the PTBL, MPHN and CPZ groups.

Table 30. LH release and ovulation induced by electrical stimulation of the MPOA in rats blocked with PTBL, MPHN, CPZ, or ATR plus PTBL. (From Everett and Tyrey 1982b)

Blocking agent	No. of rats	Serum LH (ng/ml)[a]			No. of tubal ova
		0 time[b]	60 min	90 min	
PTBL	7	60±5	430±42	306±30	9–16
MPHN	5		571±132	683±195	12–15[c]
MPHN (sham)	9	95±16		76±6	0
CPZ	5	53±3	483±136	434±90	11–14
CPZ (sham)	9	62±5	57±5	55±5	0
ATR plus PTBL	7		310±74	271±79	

[a] Expressed as mean ±SEM.

[b] Sampled under ether in late morning or early afternoon, excepting some MPHN-blocked rats sampled just before stereotaxis.

[c] One rat failed to ovulate and had only two follicles with prelutein activation.

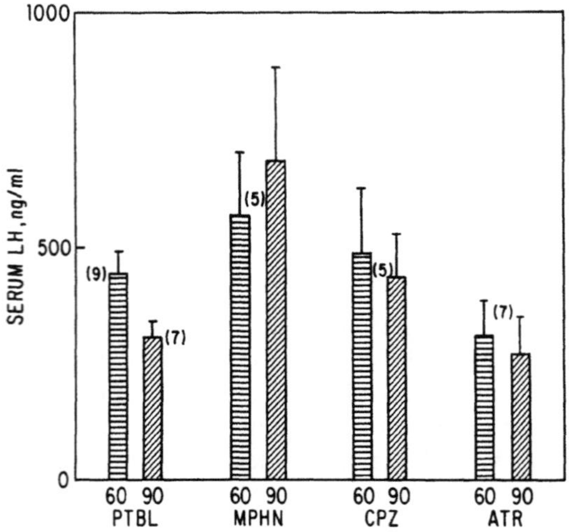

Fig. 55. Serum LH concentrations 60 and 90 min after the start of electrical stimulation of the MPOA in proestrous rats blocked with pentobarbital (PTBL), morphine (MPHN), chlorpromazine (CPZ), or atropine (ATR) supplemented with pentobarbital. *Parameters*: four 5-min bursts of 750 μA pulse pairs at 30 Hz, on and off each 15 sec for 60 min. Each bar represents the mean with its SE above. Only the 90-min values for MPHN and ATR are significantly different ($P < 0.05$). (From Everett 1982b)

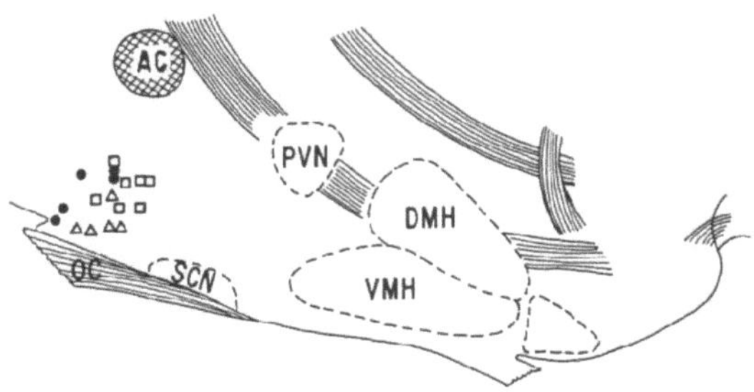

Fig. 56. Approximate electrode sites for electrical stimulation of the POA in proestrous CD rats blocked with MPHN (●), CPZ (Δ) or ATR (□), projected on a parasagittal diagram of the hypothalamus. Abbreviations as in Fig. 53. (From Everett and Tyrey 1982b)

The salient contribution of this study is the demonstration that the LH-release mechanism remains responsive in spite of the blocking action of any of the four dissimilar drugs. Urethane and Δ^9-tetrahydrocannabinol (THC) should be added to the list. As mentioned earlier, Sherwood et al. (1980) described surges of LH induced by MPOA electrical stimulation in rats during prolonged anesthesia with urethane. Murphy and Tyrey (1986) reported the induction of ovulatory surges of LH by ECS of the MPOA in rats blocked with THC. None of the several drugs has its primary ovulation-blocking action on the preoptic-tuberal system, on the availability LHRH or its release into the pituitary-portal vessels, on the capacity of the pituitary to release LH, or on the responsiveness of the ovaries to the LH surge.

Stimulation of the Preoptic-Tuberal System During Persistent Estrus

Clemens et al. (1969) and Terasawa et al. (1969) concurrently reported that ovulation can be induced during SPE by preoptic ECS. I had also obtained that result in 9/13 O-M rats during 1962–1967 (unpublished), 4 rats yielding full ovulation of 10–15 ova. Wuttke and Meites (1973) reported a moderate increase of serum LH concentration 45 min after bilateral ECS of the MPOA in 20- to 24-month old multiparous SPE rats; the level was diminishing at 2 hours and at baseline by 3 hours. Ovulation occurred in 6/9 rats.

In our first report (Everett et al. 1970) we described differences between O-M and CD rats with respect to preoptic ECS not only in cyclic rats, but also during SPE and the persistent estrus of androgenized rats (TPPE) (Table 31). The O-M SPE rats of that study were virgin females aged 12–19 months. The relatively early onset of SPE in CD virgin females later gave us a ready source of middle-aged SPE rats that could be compared with cycling rats of similar age (8½ to 11 months). As shown in Table 31, 6 O-M SPE rats were unilaterally stimulated with 100 μA D.C. × 30 sec. Three ovulated completely (10–12 tubal ova) and another was probably complete, having 5 corpora lutea in the left ovary during the subsequent pseudopregnancy. The fifth rat produced only 3 ova and the sixth failed to respond at all, presenting a set of atretic follicles several days later. CD SPE rats, by contrast, were highly refractory to the unilateral stimulation. Three showed prelutein changes in some follicles, but 3 others failed to show even histologic signs of ovarian activation. Somewhat better response was obtained with bilateral ECS in 10 other CD rats. This double stimulus

Table 31. Strain differences for preoptic electrochemical stimulation of ovulation in spontaneous and androgen-induced persistent estrus. (Data from Everett et al. 1970)

Strain	Status	ECS	μCoulombs	Results
O-M	SPE	100 μA × 30s	3000	+ + + + p o
CD	SPE	100 μA × 30s	3000	p p p o o o
		100 μA × 30s bilateral	6000	+ + + p p p p p o
O-M	TPPE	100 μA × 30s bilateral	6000	+ + + +
		10 μA × 20s	200	p p
		10 μA × 40s	400	p p
		10 μA × 60s	600	+ +
CD	TPPE	100 μA × 30s bilateral	6000	p o o o o o

Key: + = full ovulation.
p = small number of tubal ova or only histologically detectable lutein changes.
o = Neither tubal ova nor lutein changes.

ovulated 3 rats, produced moderate prelutein changes in some follicles of 5 rats, and was entirely negative in only 2 rats.

Similar strain differences were encountered in neonatally andogenized (TPEE) rats. Testosterone propionate (1.25 mg) had been injected subcutaneously postnatally on day 5. At the time of experimentation 8 mo later all presented constant vaginal estrus and were each subjected to bilateral MPOA ECS with 100 μA × 30 sec (6000 μcoulombs). Whereas all 4 O-M rats ovulated full sets of 8 to 11 ova, only 1/6 CD rats ovulated (5 ruptured follicles plus many luteinizing). The other 5 CD rats gave no histologic sign whatever of follicle activation. To test the MPOA threshold in O-M TPPE, six additional rats were unilaterally stimulated at lower intensity (10 μA × 20–60 sec: 200–600 μcoulombs). All ovulated, but only the 2 rats receiving 600 μcoulombs showed full ovulation. Although the threshold appeared to be somewhat higher than in proestrous O-M rats, it was clearly well below that in CD TPPE animals. Barraclough and Gorski (1961) and Gorski and Barraclough (1963) had previously reported failure of MPOA stimulation to induce ovulation in TPPE Sprague-Dawley rats unlehs they were primed with progesterone. Those parameters of stimulation (80–100 μA, 0.5 msec rectangular monophasic pulses, 100 Hz, 15 sec "on" and 15 sec "off", for 15 min) would have introduced 3600 to 4500 μcoulombs, somewhat less than our larger stimulations, but far above the O-M treshold. It is appropriate to recall that CD rats are Sprague-Dawley derivatives. It is also appropriate to note that Quinn (1966), utilizing castrated male O-M rats bearing ovarian grafts, induced corpus luteum formation in 11 of 12 rats receiving MPOA ECS with μA for 40–60 sec. Even 10μA for 20 sec luteinized 4 to 8 rats. Thus the threshold was of the same order as that in O-M TPPE females. The report of similar experiments by Moll and Zeilmaker (1966) is difficult to evaluate in the present context.

Further investigation of CD SPE rats (Everett and Tyrey 1983) was pursued by *electrical* stimulation of the MPOA and ARC/ME in experiments concurrent with those in proestrous rats described on pp. 80–82. The electrical parameters were like those in the cyclic rats: 750μA peak-to-peak in the MPOA and 500μA in the ARC/ME. In Series I, MPOA stimulation extended for 60 min and ARC/ME stimulation for 45 min. Blood samples for LH radioimmunoassay were obtained at 60 min from an external jugular vein and at 90 min by cardiac puncture. In Series II all rats were bled sequentially from intracardiac catheters installed between 0900h and 1100h (ether anesthesia). All subjects were anesthetized with PTBL (3.5 mg/100 g BW) injected shortly before stereotaxis in the early afternoon.

In *Series I* (Table 32) there was little difference between SPE rats and proestrous rats in their serum LH concentrations at 60 min and 90 min. In the ARC/ME group the concentrations were somewhat higher than in the MPOA group although not statistically significant.

Series II. Because the LH levels in Series I were much lower than those expected in the normal proestrous surge and we knew that normal levels could be reached in proestrous rats by extending the stimulation time to 2 hours, the effect of prolonged stimulation was examined in SPE rats. All were from one shipment and of comparable age (Table 32, Fig. 57). Only four proestrous cycling rats of that shipment were available at the time of the study and they were subjected to MPOA stimulation. The SPE results are of primary interest in showing that during the second hour of stimulation the serum LH continued to rise and that its rate of increase was essentially alike in the MPOA and ARC/ME groups (slopes between 60 and 120 min: 8.50 and 7.05, respectively). Although as in Series I there was no firm difference between the two groups at 60 min, the ARC/ME data overall showed a highly significant advantage ($P < 0.01$). In spite of the continued rise during the second hour, the LH values on the average remained well below the amounts

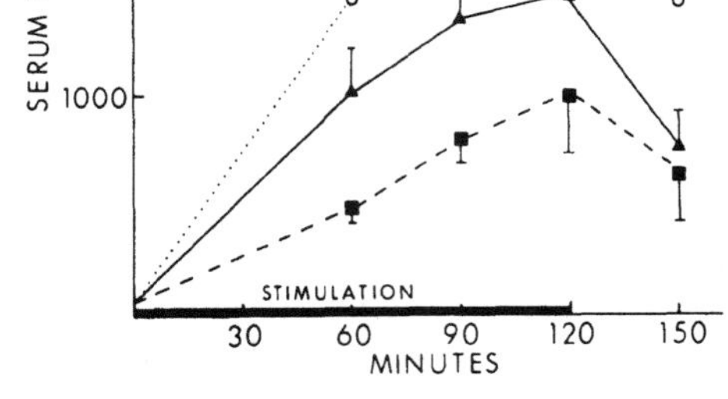

Fig. 57. Comparative patterns of LH secretion during and after electrical stimulation of MPOA and ARC/ME sites in spontaneously persistent estrous and proestrous cyclic CD rats. (From Everett and Tyrey 1983)

Table 32. Surges of LH induced by electrical stimulation of the MPOA or the ARC/ME in SPE rats compared with proestrus rats.
(From Everett and Tyrey, 1983)

Reproductive Status		Age range (days)	Stimulation[a]		No. of rats	Serum LH (ng/ml)[b]			
			Site	Duration		60 min[c]	90 min	120 min	150 min
Series I.	SPE	210–354	MPOA	60 min	9	428 ± 74	432 ±57		
	Cyclic, proestrous[d]	172–263	MPOA	60 min	7	498 ± 88	422 ±62		
	SPE	225–265	ARC/ME	45 min	4	681 ±275	449 ±224		
	SPE	174–315	ARC/ME	45 min	7		465 ±202		
	Cyclic, proestrous[d]	129–173	ARC/ME	45 min	10	670 ±114	457 ±58		
Series II.	SPE	193–229	MPOA	120 min	5	480 ± 59	795 ±110	990 ±257	614 ±199
	SPE	212–272	ARC/ME	120 min	7	1012 ±194	1332 ±250	1435 ±292	762 ±148
	Cyclic, proestrous	233–236	MPOA	120 min	4	1433 ±220	2116 ±350	2394 ±486	1410 ±322

[a] Biphasic pairs of 1-msec pulses, 30 Hz, 30/60 sec; 750 μA peak-to-peak for MPOA, 50 μA for ARC/ME.
[b] In terms of NIADDK rat LH RP-1. Mean ±SEM.
[c] Time after beginning stimulation.
[d] Data previously published from experiments concurrent with most of the SPE experiments of this series.

attained in proestrous rats. Only 5 of the 12 SPE rats produced peak levels above 1500 ng/ml. Ovarian responses, even to the sustained levels of Series II, were more erratic than in proestrous rats similarly stimulated. This was ascribed to variations in number of competent follicles.

Thus, during the various types of persistent estrus (SPE, LLPE, and TPPE) the absent ovulatory surge of LH can be induced experimentally by brain stimulation. Castrated male rats bearing ovarian grafts may similarly be induced to luteinize. Furthermore, during pseudopregnancy, another condition in which ovulation is inhibited, preoptic ECS can induce it (Everett, unpublished). Seventeen 4-day cyclic

O-M rats were cervically stimulated electrically on the morning of estrus. During the following prolonged diestrus, all were subjected to bilateral MPOA stimulation with 100 μA D.C. × 30 sec. On the next day, from 1 to 9 tubal ova were present in all cases; the corpora lutea of pseudopregnancy were strongly fatty and positive for cholesterol in the Schultz test. Stimulation had been applied on day 3 of diestrus (5 rats), day 5 (6 rats), day 8 (5 rats) and day 9 (1 rat).

Overview of the Stimulation Studies

1. The natural ovulatory surge of LH can be reproduced by stimulation of a septal-preoptic-tuberal system (SPTS) which is rostrally diffuse and caudally convergent upon the MBT.
2. The amount of LH discharged is quantitatively dependent on the extent of involvement of the SPTS and the duration of the stimulus. With ECS, to discharge an ovulatory quota of LH within 30 min requires bilateral involvement of most of the SPTS. Smaller, unilateral ECS lesions can produce the quota during a longer time. The ovarian response to LH depends on both the level of the circulating hormone and its time of action.
3. The need for greater SPTS stimulation in late diestrus reflects in part a lower ovarian responsiveness to LH than in proestrus, but the AP in late diestrus also releases less LH in response to administration of LHRH, a difference that can be erased by estrogen or progesterone priming. Whether there are also differences in the amount or patterns of LHRH release remains unknown.
4. Although ECS is produced by brief electrolysis, the irritant effect of the deposited iron lasts far longer. For several hours the neurons in the neighborhood of the initial coagulated zone remain histologically normal and presumable activated. By the next day they have disappeared, leaving an extensive halo.
5. Destructive lesions in the MBT, produced by radiofrequency current or surgical damage, can provoke LH release and ovulation.
6. Electrical stimulation of either the MPOA or the ARC/ME shows a quantitative relationship among stimulus intensity, its duration, and the amount of LH released. Presumably because of the anatomical convergence of the SPTS upon the ARC/ME, the latter region has a lower threshold for electrical stimulation than the MPOA. Although electrical stimulation of the ARC/ME might involve most of the SPTS (at least unilaterally), when applied more rostrally it probably can reach only portions of the dispersed system: hence the smaller amounts of LH released by MPOA stimulation.
7. In proestrous rats blocked with pentobarbital, electrical stimulation of either the MPOA or the ARC/ME for 2 hours produces patterns of LH release closely resembling the natural proestrus surge. Since the patterns produced by stimulation of the two regions are parallel, it seems that the SPTS operates as a unit.
8. The SPTS remains potentially functional whether prevented from spontaneous activity pharmacologically or by physiological circumstances, as in pseudopregnancy, cyclic diestrus, SPE, LLPE, TPPE, or the naturally androgenized status of the male rat.

Pseudopregnancy: Controls For Prolactin Secretion

In retrospect, the Long and Evans (1922) monograph contained the first evidence that different AP secretions are responsible in rats for ovulation and for corpus luteum function in early pregnancy or pseudopregnancy. It was almost 20 years later that Astwood (1941) purified material from rat pituitaries that was distinct from FSH and LH and had the selective power of maintaining luteal secretion. Evans et al. (1941) identified the luteotropic material as prolactin.

The choice of the rat model during the years of search for the several trophic hormones of the AP, coupled with the very fact that the corpora lutea of ovulation in that species are relatively inactive during the foreshortened estrous cycles, led inevitably to considerable confusion and misinterpretation of experimental data. It was common to regard the luteal phase of the mammalian cycle as the natural sequel to ovulation and structural transformation of the Graafian follicles to corpora lutea, as if the controlling mechanisms were essentially the same. In most mammals, both ovulation and the period of luteal function were seen to be spontaneous. Even in the special cases of "reflex" ovulation like the rabbit and the cat, the coital stimulus evoked first ovulation and formation of corpora lutea, after which came pseudopregnancy of several weeks' duration. Rats, mice and hamsters, however, present the ovulation phase spontaneously, while activation of the corpora lutea for pseudopregnancy or the progravid stage of pregnancy requires genital stimulation. There was a tendency to regard the sexually provoked ovulation in rabbits and pseudopregnancy in rats as involving the same mechanisms.

Thus, Taubenhaus and Soskin (1941) searching for neurohumoral controls of the release of "luteinizing hormone" exposed the pituitary gland by the retropharyngeal route, bathed the gland by one or another drug, and measured the effect by the occurrence or absence of pseudopregnancy. In a series of studies by Westman and Jacobsohn (1936–1942) it was consistently assumed that ovulation, luteinization and corpus luteum activation are sequential results of the same process.

Westman and Jacobsohn demonstrated unequivocally that cutting the rabbit pituitary stalk prevented ovulation in response to mating. Likewise, in prepubertal rats, they showed that, while administration of estradiol benzoate induced ovulation, that process could be interrupted by cutting the pituitary stalk one or two days after injection. In adult rats, on the other hand, they obtained ambiguous results with stalk sectioning. The animal was first stalk-sectioned and then the cervix was stimulated; the resulting pseudopregnancy was interpreted to mean that the cervical stimulation had released gonadotropin which caused both ovulation and corpus luteum function. It was unfortunate that there were no control experiments omitting the cervical stimulation, for in retrospect the result was really the first indication that the hypothalamus tends to restrain prolactin secretion. Similar information was lacking in experiments by Desclin (1950), testing for pseudopregnancy after transplanting the AP to the kidney. In both intact rats and those bearing the transplants, implantation of a pellet of estradiol resulted in pseudopregnancy, recognized by mucification of the vaginal epithelium. Considering both studies, it seemed probable that merely to disconnect the AP from the brain would be sufficient to produce pseudopregnancy.

Removal of the AP from Direct Influence of the Hypothalamus Favors Secretion of Prolactin

This was fully demonstrated in several studies from this laboratory (Everett 1954, 1956; Nikitovitch-Winer and Everett 1958a,b, 1959).

Short-term Experiments (Everett 1954)

Thirty adult female rats of the O-M strain were hypophysectomized by the parapharyngeal route on the day after ovulation. The AP was autotransplanted in 12 of the rats to a pocket under the renal capsule and in 2 rats deeply into the neck. The other 16 rats served as controls without grafting, nine being completely or nearly completely hypophysectomized and seven retaining extensive fragments of the AP. The completeness of hypophysectomy was assessed histologically in serial sections of the cranial floor through the original site of the gland. The left uterus was traumatized 4 days after hypophysectomy and the effect was assessed at 8 days (Table 33). Deciduomata were present in all rats bearing AP grafts and in 6 of the 7 control rats retaining extensive AP fragments in the original site. The nine hypophysectomized controls, complete or almost complete, showed no decidual response whatever.

Table 33. Deciduoma formation in female rats bearing autografts of anterior hypophysis. (From Everett 1954)

Group no.	Hypophysectomy	Grafted to:	No. of rats	Deciduoma reaction ++	+++	++++	Total pos.	Total neg.
I	Complete	–	3				0	3
II	Complete	Renal capsule	6		1	5	6	0
III	Nearly complete	–	6				0	6
IV	Nearly complete	Renal capsule	6		2	4	6	0
		Neck	2		1	1	2	0
V	Extensive fragment left	–	7	2	1	3	6	1

Long-term Experiments

Although the short-term experiments satisfied the hypothesis that AP grafts isolated from the brain produce enough prolactin to maintain luteal function, there was a remote possibility that the luteotropic function did not represent true secretion or that some irritative effect at the time of the transplantation caused a transient period of secretion. In testing these questions (Everett 1956), 24 rats on the day after ovulation were subjected to hypophysectomy followed by AP transplantation to the kidney. The test for pseudopregnancy in these cases was vaginal mucification and lack of cornification after massive treatment with 125 μg estradiol benzoate 3 times daily for 5 days. The experiments and results are summarized in Table 34; sections of representative ovaries are shown in Fig. 58. Even 3 months or more after the transplantations the vaginas were consistently mucified by estrogen treatment, except in Group C after the AP graft had been removed. One week after that the corpora lutea had markedly regressed and the vaginas were heavily cornified in response to a new test with the estrogen; previous tests while the grafts were still in place had uniformly shown mucification.

Beyond question, therefore, the grafts secreted prolactin and continued to do so for many weeks without interruption. As shown in Fig. 60 there was a uniform absence of growing follicles, indicating deficient FSH secretion. The interstitial tissue showed the "wheel nuclei" and reduced cytoplasm characteristic of long-standing LH deficiency. There were also deficiencies in TSH and ACTH as shown by atrophy of the thyroids and the adrenal cortex. Thus, it was clear that removal from the hypothalamic influence favored the continuing secretion of prolactin while other trophic secretions were greatly diminished or lost.

Table 34. Long-term experiments with pars distalis autografts on the kidney. (From Everett 1956a)

Group	No. of animals	Duration in days[a]					Estrogen test	Vaginal response
A	6	**26**[b]	53[b]	53[b]			None	————
		64	64	**90**				
B	13	23	**33**	33	33	50	Final Week	Mucification
		60	60	60	88	**89**		
		90	104	104				
C	5	**94**[c]	94[c]	**98**[d]	**103**[d]		Before graft removal	Mucification
		120[d]						
							After graft removal	Cornification

[a] Figures in bold face represent animals in which no trace of pars distalis remained in the hypophyseal capsule. Others retained small fragments of doubtful significance.

[b] Dates of castration, after which the standard estrogen regime was instituted, with resulting cornification of the vagina. Autopsy was 7 days after castration.

[c] Tested with estrogen on 3 occasions before graft removal at 88 days (see text).

[d] Tested with estrogen on one occasion shortly before graft removal.

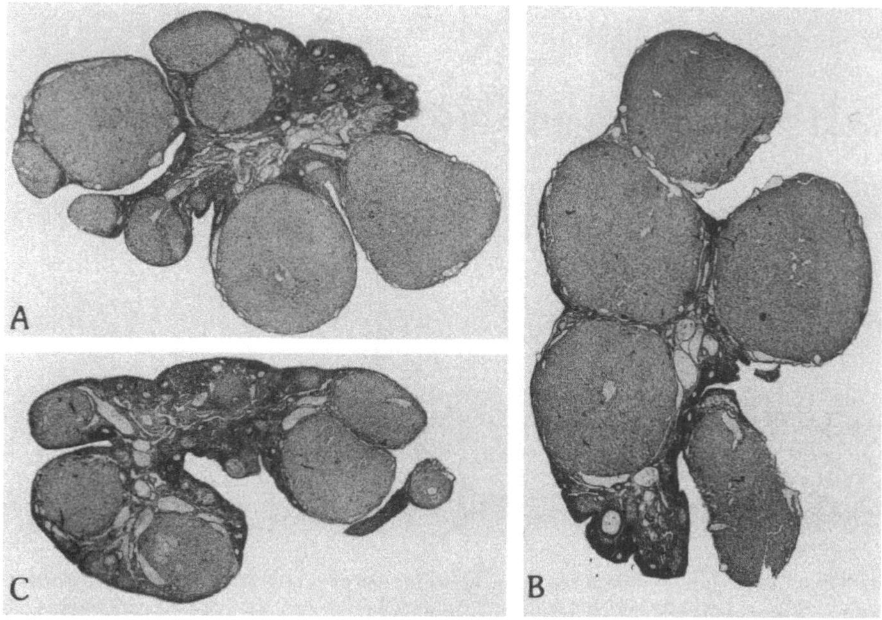

Fig. 58. Ovarian histology of the three experimental groups in the long-term series of pituitary transplantation to the kidney presented in Table 34. Note the atrophy of interstitial tissue and follicular apparatus in all examples. **A** Untreated, 90 days after transplantation. Three primary and several secondary corpora lutea. **B** Enlarged corpora lutea in a 60-day experiment after 125 μg estradiol benzoate during the final week. **C** Corpus luteum atrophy 8 days after removal of the graft in a 94-day experiment, 125 μg estradiol benzoate during final week. (From Everett 1956a)

An ovarian feature not anticipated was the retention of secondary, smaller corpora lutea that were judged to represent corpora lutea from previous cycles whose atresia was interrupted by the AP transplantation (see Fig. 58A). Cells of the secondary corpora lutea appeared identical to those of the primary set and the tissue was as well organized. In the ovaries of Group A, diameters of the primary corpora lutea were approximately 1.5 mm like those of normal pseudopregnancy. In samples removed after estrogen treatment (Group B) the diameters had increased to ~2.0 mm as in corpora lutea of advanced pregnancy. This reflected the enlargement of the individual cells; cells of the secondary corpora lutea were correspondingly enlarged, also suggesting functional status.

Further long-term experiments (Nikitovitch-Winer and Everett 1958a) addressed the possible effects of stage of the cycle at the time of transplantation and of the site where the graft was implanted. In that work a few initial trials showed that the decidual reaction can be obtained weeks after the transplantation; consequently this was the test chosen for luteotropic secretion. Table 35 shows the results of grafting on different days of the cycle. Only in proestrus were the corpora lutea unresponsive

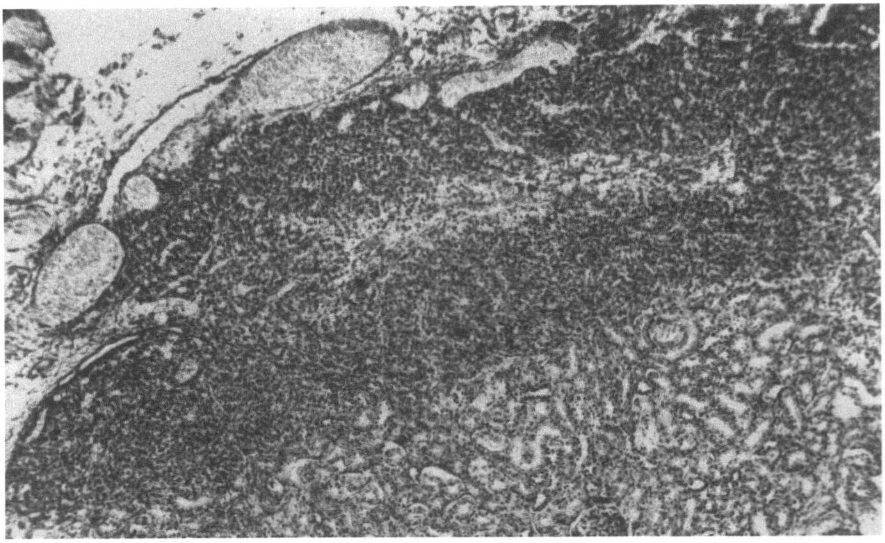

Fig. 59. A pars distalis graft removed from the kidney 86 days after transplantation. A portion of renal parenchyma is shown in immediate contact with the graft. Note the prominent veins external to the graft. (From Everett 1956a)

(presumably the few exceptional rats were not truly in proestrus). Foreshadowing the lack of response in proestrus, the 9 rats operated upon during D-3 of the 5-day cycle presented corpora lutea that were noticeably smaller than after operations during estrus and D-2; only three of the nine gave maximal decidual reactions. To test the luteotropic power of the grafts in the proestrus group, new corpora lutea were induced by gonadotropin administration (FSH for 3 days, then LH on the following day) beginning 23 to 67 days after transplantation. Whereas all rats formed no deciduomata in the first tests, the 8 rats tested later gave a moderate-to-maximal response. Therefore, the stage of the cycle had no evident effect on luteotropic power of the graft.

Since the anterior chamber of the eye has been a favorite site for experimental organ transplantation, this site was the one chosen for comparison with the effectiveness of the kidney capsule location, although some information was available from the grafts placed in fascia of the neck and from a few cases in which the graft slipped out of the renal capsule into the neighboring connective tissue. The

Fig. 60. Ovaries of (**A**) a normal O-M cycling rat in proestrus, (**B**) a cycling experimental rat 98 days after re-transplantation of the pars distalis from the kidney to the median eminence, (**C**) an anestrous rat after re-transplantation from the kidney under the temporal lobe, and (**D**) an anestrous rat after transplantation to the kidney during proestrus when the corpora lutea have lost ability to respond to prolactin. (From Nikitovitch-Winer and Everett 1958b)

Table 35. Pituitary transplantation to the kidney capsule at different stages of the estrous cycle. (From Nikitovitch-Winer and Everett, 1958a)

No. of rats	Stage of cycle	Interval to 1st trauma (days)	Decidual response	Interval to 2nd trauma (days)	Decidual response
2	Estrus	8,13	+ + + +	22–37	+ + + +
7	D-2	3–13	+ + + to + + + +	17–37	+ + to + + + +
9	D-3[a]	4–10	+ + to + + + +		
2	Proestrus	4	0		
2	Proestrus	6,7	0	41–54[b]	0
8	Proestrus	5–21	0	29–58[b]	+ + to + + + +
3	Proestrus	7–15	+ + to + + +		

[a] 5-day cycle.
[b] Before 2nd trauma a course of FSH-LH treatment was given to produce new corpora lutea.

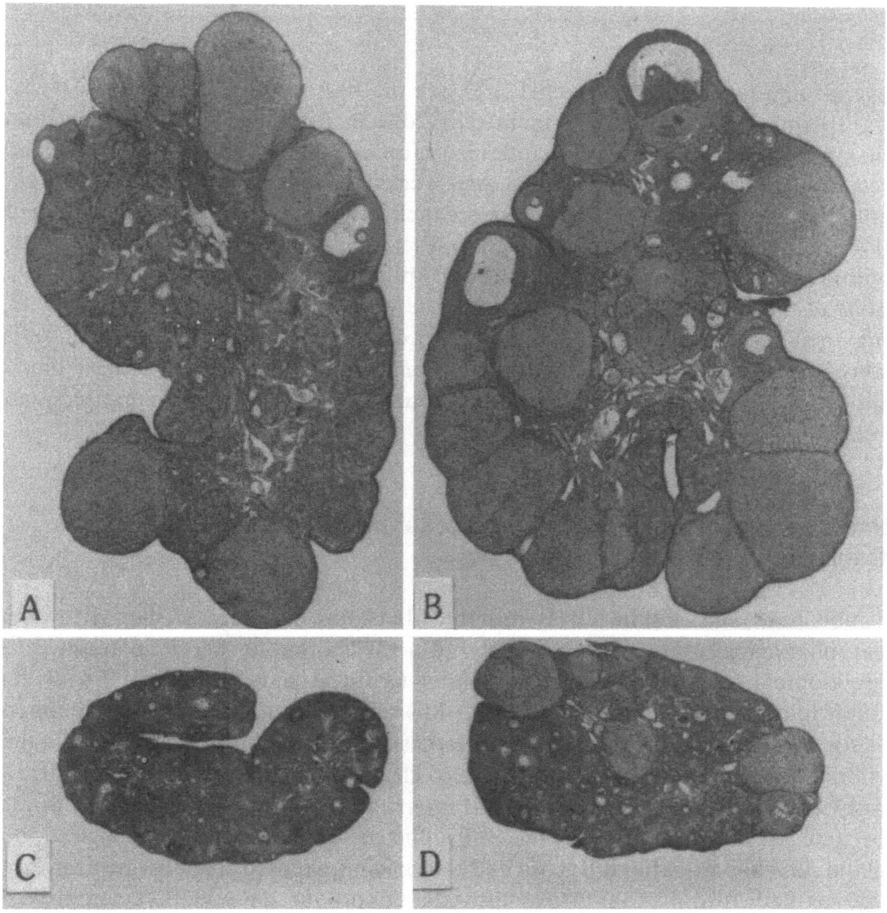

Table 36. Pituitary autografts to the anterior chamber of the eye (Series I). (From Nikitovitch-Winer and Everett, 1958a)

A. Animals with functional grafts					*B. Animals with non-functional grafts*		
Animal No.	Days operation to 1st trauma	Decidual response	Days, operation to 2nd trauma	Decidual response	Animal No.	Days, operation to trauma	Decidual response
1	12	+ + + +	40	+ + + +	1	5	0
2	10	+ + + +	31	+ + + +	2	12	0
3	5	+ + + +	42	+ + & + + +	3	12	0
4	6	+ + + +	36	+ +	4	13	0
5	10	+ + + +	31	+ +	5	10	0
6	13	+ + + +	41	0			
7	5	+ + + +	42	0			
8	10	+ + + +	36	0			
9	9	+ + + +	70	0			

overall finding was that the renal site gave the best results. The results of autotransplantation to the eye are presented in Table 36. Although 9 of 14 rats presented maximal decidual reactions to uterine trauma 5–13 days after the operation, responsiveness to the second trauma some weeks later was usually reduced or absent entirely. The grafts in 5 of the 14 rats were ineffective from the start; examination of the eye at the time of necropsy failed to disclose glandular tissue. Whereas grafts within the kidney capsule were consistently plump and well vascularized (Fig. 59), none of those in other sites was as large and well organized, even in rats showing maximal decidual reactions on successive occasions. Interestingly, although grafts were best organized when in contact with renal parenchyma, their principal blood supply eventually came from extrarenal sources, typically the ovarian artery and vein.

Pseudopregnancy Induced by Damage to the AP, Stalk-Section, or Damage to the Basal Hypothalamus

Taubenhaus and Soskin (1941) reported that after exposure of the AP by the parapharyngeal approach, application of acetylcholine to the gland resulted in pseudopregnancy. In experiments intended to repeat that experiment, I made the following observations (unpublished). O-M 4-day cyclic rats in estrus were prepared as for hypophysectomy under ether anesthesia, the usual hole being drilled in the cranial floor. In 7 rats the AP was exposed cleanly in the depths of this hole. In all cases estrous cycles continued without interruption. On the other hand, 3/4 rats became pseudopregnant after the drill inflicted slight superficial damage to the gland. In other rats, after the gland had been cleanly exposed, a deliberate tranverse cut was made rostrally so as to sever the stalk; this produced pseudopregnancy in 4/5 rats. Comparable results from stalk sectioning were obtained by Nikitovitch-Winer (1957, 1965). Among 9 rats, in which the stalks were cut during estrus by the transtemporal approach without placement of effective barriers to vascular regeneration, all experienced pseudopregnancy-like delays of renewed cycling; one rat was appropriately tested by uterine trauma on day 4 after operation, producing a strong decidual response. Effective barriers were placed in 8 estrous rats. All displayed permanent anestrus; prolonged corpus luteum function was demonstrated by the decidual reaction to trauma, first at 4–36 days after operation and by repetition in 6 of the 8 rats at 46–74 days.

It is important to note that this last experiment was essentially a repetition of the Westman-Jacobsohn experiment cited earlier, but now without stimulation of the vaginal cervix. However, there may have been an inadvertent "non-specific" stimulus resulting from the mechanical manipulation of the median eminence at the time of the operation. Nikitovitch-Winer discovered that when 11 estrous rats were thus treated without damage to the stalk itself all became pseudopregnant. By contrast, only 2/12 rats showed any interruption of cycling after the brain was exposed and the temporal lobe lifted without disturbing the basal diencephalon. Thus, any slight damage to either the median eminence or the AP, or the temporary interruption of the hypophysial portal vessels can institute pseudopregnancy. Permanent interruption has the same effect as transplantation of the AP to a distant site.

Re-Transplantation of AP Grafts from Kidney to Brain

These investigations (Nikitovitch-Winer and Everett 1957, 1958, 1959) were based on three earlier studies by Greep (1936), Harris (1950) and Harris and Jacobsohn (1952). Greep demonstrated that after hypophysectomy, if autografts or homografts were made into the original site, estrous cycles were sustained, pregnancy and lactation were possible, and the animals grew well. Harris, by demonstrating the capacity of the hypophysial portal vessels to regenerate, showed the recovery of AP functions to be directly related to the degree of this vascular regeneration. Harris and Jacobsohn, immediately after hypophysectomy, homografted the AP intracranially via the transtemporal approach to a location near the median eminence or (as controls) under the temporal lobe. The subjects were post partum and most of the grafts were multiple pituitaries from the animals' own offspring (age: 1–10 days). In a few cases hypophysectomized female subjects received pituitaries from adult male donors. Estrous cycles reappeared consistently in the rats bearing grafts near the median eminence, several pregnancies were obtained, and normal young were delivered at term. The grafts from the new-born donors acquired adult functions far in advance of their normal time. Furthermore, male grafts in female hosts readily acquired the female pattern of AP secretion.

With the Harris-Jacobsohn study in mind, it occurred to us (Nikitovitch-Winer and Everett 1957ff) that if AP autografts were left on the kidney for several weeks, long enough to lose gonadotrophic actions, replacement near the median eminence could restore estrous cycles and related reproductive functions. Such was the case.

The experiments were carried out in 62 female O-M rats 3½ to 7 months old and in proestrus at the start. In addition 10 normal controls were neither hypophysectomized nor engrafted with AP tissue. The 62 rats were hypophysectomized by the parapharyngeal route and the AP autografted to the kidney capsule in routine manner. In the definitive experimental group the grafts remained on the kidney 29 to 33 days, except in one case for only 21 days. The grafts were then removed and re-transplanted under the median eminence by Harris' transtemporal approach (Group MEm-1, 14 rats). In Group MEm-2 (10 rats) the grafts were on the kidney 15 to 19 days before re-transplantation. These rats were briefly treated with FSH and LH 7 to 28 days later in the mistaken hope of aiding ovarian recovery. In control Group TL (21 rats), after being on the kidney for 12 to 26 days, the grafts were re-implanted under the temporal lobe of the brain. In control Group Ky (21 rats) the grafts merely remained on the kidney.

Vaginal smears were prepared during the postoperative period six days each week. At necropsy the ovaries, thyroids and adrenals were trimmed, weighed and preserved in Bouin's fluid, together with portions of uterus and vagina. Parts of the cranial floor including the hypophysial capsule were decalcified and preserved to be serially sectioned and searched for remnants of the gland in the original site. The routine stain was a modified Mallory trichrome stain. For cytologic study, all grafts were fixed in Elftman's chrome alum solution at least 5 hours and washed overnight. Alternate strips of 5 μm paraffin sections were stained with the periodic-acid Schiff (PAS) method (Purves and Griesbach 1954) or Halmi's (1952) aldehyde fuchsin (AF) trichrome method. The location of grafts near the median eminence was charted and the adjacent portion of the hypothalamus left attached to the grafts.

In Group MEm-1 all but one of the 14 rats began to cycle spontaneously 8 to 68 days after the re-transplantation. Each was tested for fertility with a known potent male at the approach of estrus on one or more occasion. Seven became pregnant: three delivered spontaneously at 23 days, and the others were delivered artificially near term. The numbers of offspring were all below normal, ranging from 1 to 6. Although the mammary glands were well developed, none of the mothers could nurse her young.

The results in Group MEm-2 were less satisfactory, although evidence of returning gonadotropin secretion was apparent in the vaginal smears of 7 rats, beginning 11 to 66 days after the re-transplantations. Only one rat presented fairly regular estrous cycles, became pregnant after 4 trials, but later resorbed the litter. Four became persistent-estrous soon after the FSH-LH treatment.

All animals in Group Ky and Group TL remained anestrous throughout, while the normal Group N continued to cycle regularly.

Histology of the ovaries, vaginas and uteri gave further evidence of gonadotropic stimulation in the rats bearing grafts re-vascularized from the median eminence. Ovarian histology varied in typical manner with stage of the vaginal cycle. The numbers of growing and Graafian follicles were low, yet structurally no different than normal (Fig. 60). Whereas in Groups Ky and TL the ovarian interstitial tissue showed the characteristic "wheel" nuclei marking LH deficiency, in the cycling or post partum, rats of groups MEm-1 and MEm-2 there was good interstitial cell repair (Fig. 61).

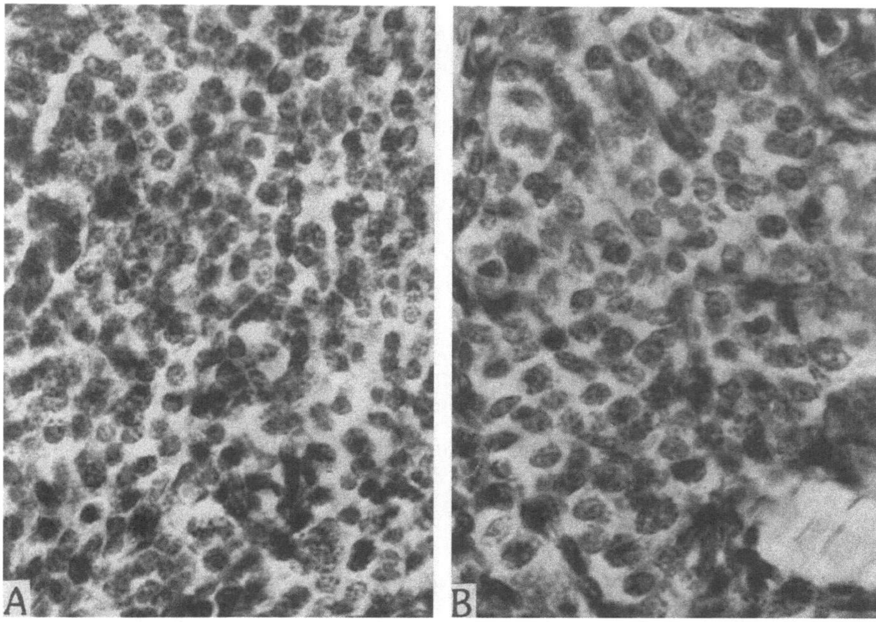

Fig. 61. **A** Interstitial tissue from the atrophic ovary shown in Fig. 60D, presenting marked LH deficiency. **B** Interstitial tissue from the ovary in Fig. 60B, showing good repair by LH from the re-transplanted pituitary. (From Nikitovitch-Winer and Everett 1958b)

Significant recovery of ACTH and TSH secretion was also evident in rats bearing grafts under the median eminence, as shown by enlargement of the adrenal cortex and the thyroids.

The histology and cytology of the grafts are of special interest. The initial insult of transplantation to the kidney left by the next day only a thin zone of healthy parenchyma near the surface (Fig. 62). This tissue retained its organization as anastomosing cell cords among well-filled sinusoids. Gonadotropic and thyrotropic basophils remained abundant though already smaller than normal. During the next few days the infarcted zone was greatly reduced while the healthy shell of parenchyma increased in thickness (Fig. 62c), partly an adjustment to the reduced circumference. There was also extensive proliferation of the parenchyma during the first week, as shown by the presence of numerous mitotic figures. By the end of the week the infarcted zone was represented by only a central scar, all parenchymal cells of the graft were smaller than normal and most were chromophobic or weakly stainable with Orange G of the Halmi stain. Loss of cells of the basophil class was pronounced and continued through the third week (Fig. 63).

Re-transplantation of the graft into contact with the median eminence and infundibular stem resulted once more in massive central infarction, again leaving only a thin shell of recognizable parenchyma (Fig. 64a,b). One week later, reorganization was well under way (Fig. 64c) especially in parts of the graft near the

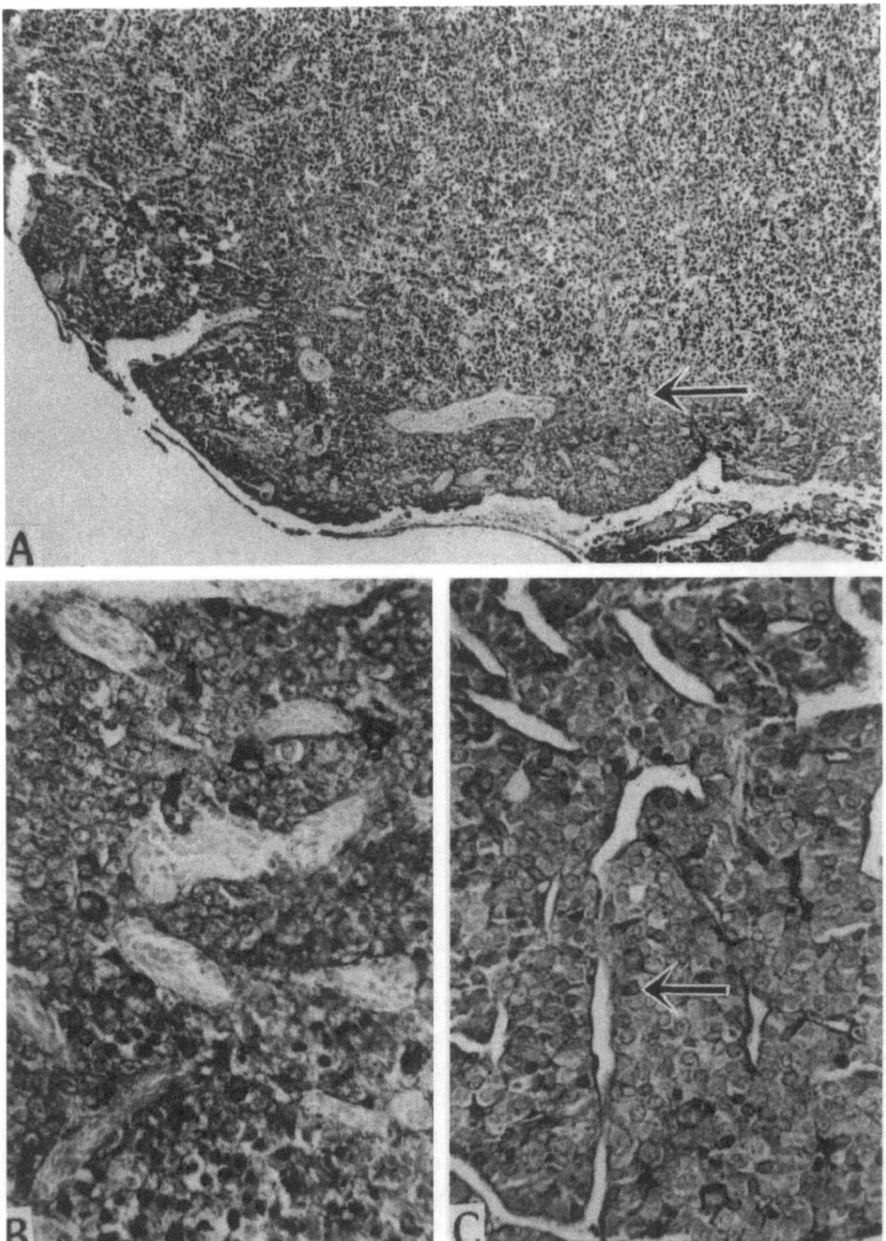

Fig. 62. Pars distalis autografts on the kidney, early changes. **A** One day after transplantation. Massive infarct in most of the interior, leaving only a thin shell of healthy parenchyma (below arrow). PAS. X100. **B** Higher magnification of peripheral zone. Free surface above. Many PAS + cells near distended sinusoids. 400X. **C** Eight days after transplantation. Most of the graft now well organized as healthy glandular tissue. Note change in the sinusoids. Arrow points to a small PAS + cell. 400X (From Nikitovitch-Winer and Everett 1959)

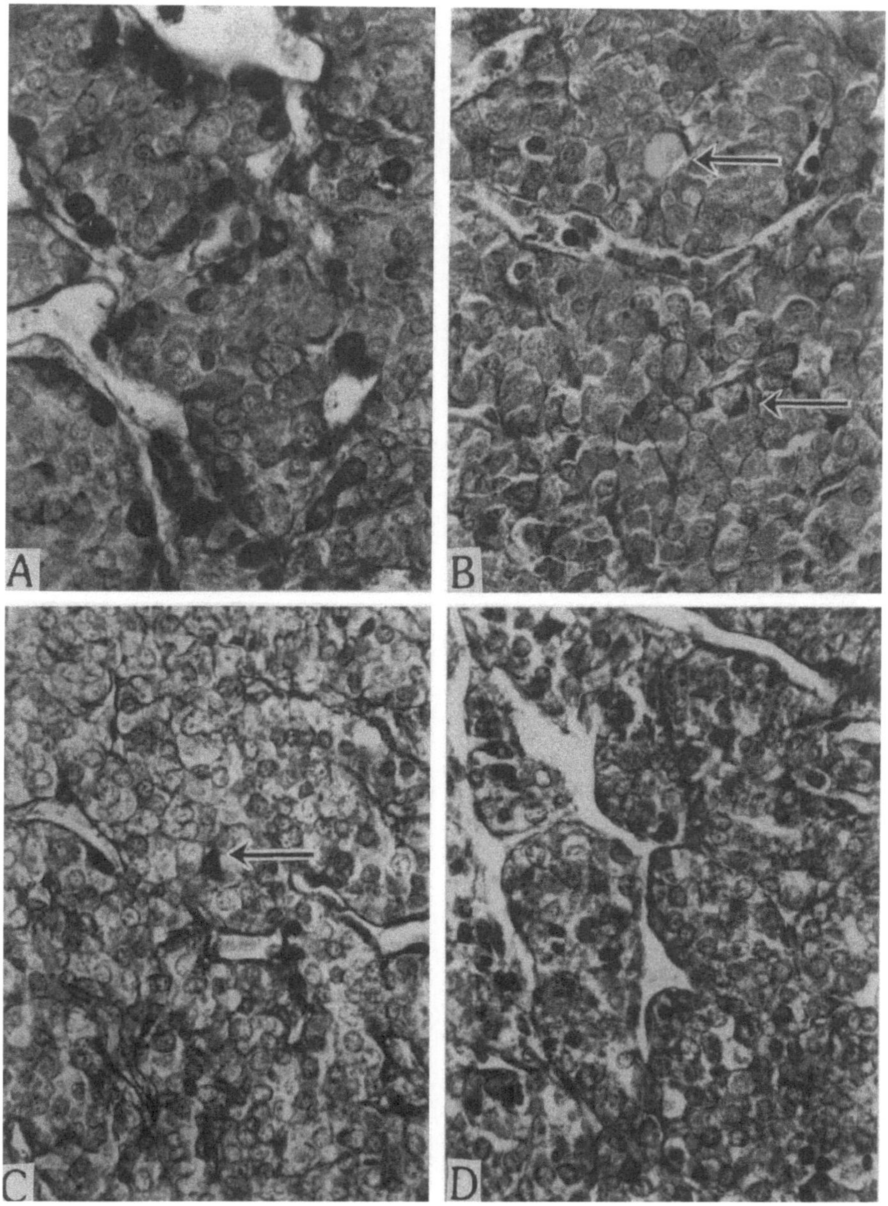

Fig. 63. Well established pars distalis grafts. 540X. **A** At 16 days after transplantation. Small gonadotrophs still frequent; especially numerous in region photographed. PAS stain. **B** At 30 days. Arrows point to a minute angular cell stained with aldehyde fuschsin (thyrotroph?) and to a large vacuole of a presumptive thyroidectomy cell. Aldehyde fuchsin stain. **C** At 94 days. Little different than at 30 days. Small PAS + cell at arrow. **D** At 40 days after re-transplantation under temporal lobe. Essentially like the grafts in B and C. PAS stain. (From Nikitovitch-Winer and Everett 1959)

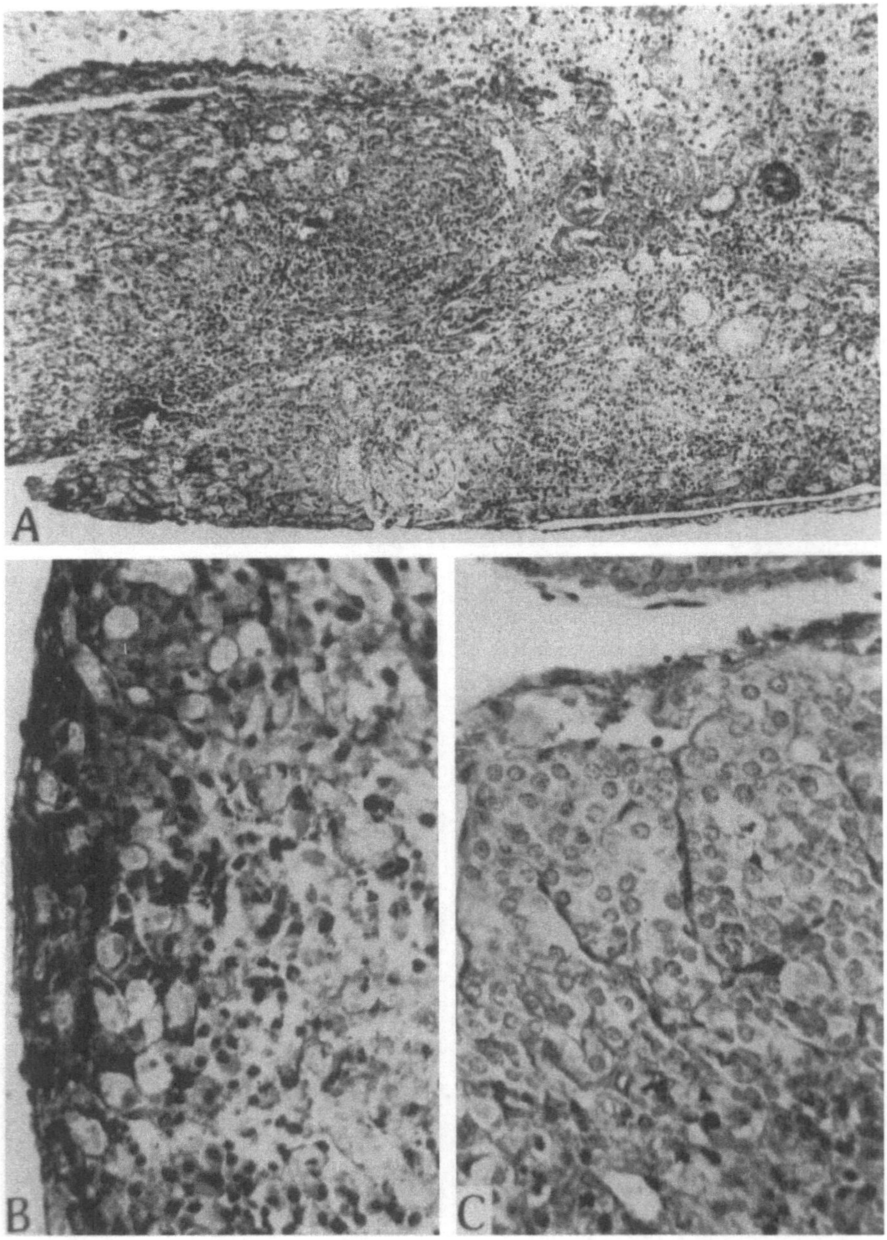

Fig. 64. Pars distalis autografts re-transplanted under median eminence. **A** One day after re-transplantation. Central infarction again leaves only a thin shell of healthy parenchyma. Union with diencephalon (above) is already intimate. PAS. 100X. **B** Higher magnification of under portion of graft in A, turned 90°. 400X. **C** Seven days after re-transplantation. A few cells of pars tuberalis above. Pars distalis parenchyma considerably increased, especially on this side near the infundibular stem. Many large, pale cells in this region; adjacent sections have mitotic figures in such cells. PAS. 400X. (From Nikitovitch-Winer and Everett 1959)

portal blood supply. Although the PAS and AF techniques failed to show any basophils at that time, there were many large cells containing the characteristic negative image of the Golgi apparatus. Specimens obtained after the animals had experienced renewal of reproductive functions, contained abundant normal or vacuolated gonadotrophs. By that time, in spite of the repeated insult, the graft had undergone remarkable recovery as well-organized glandular tissue (Fig. 65). In those animals that had only irregular cycles, the castration-cell variety of basophile was especially numerous. Thyrotrophs, stainable with both PAS and AF, were also evident, tending to be larger than normal.

Not only did the re-transplantation experiments confirm the work of Greep (1936) and of Harris and Jacobsohn (1952) in showing maintenance of cyclic gonadotropin secretion by AP grafts receiving blood from the pituitary portal vessels, but the dramatic recovery of such powers after the repeated massive infarctions at the successive operations further emphasized the importance of that blood supply. The partial recovery of ACTH and TSH secretions was also significant in that respect. Philip Smith (1961, 1963) demonstrated in both male and female rats that months after parapharyngeal hypophysectomy AP homografts inserted under the median eminence recovered not only gonadotropic, corticotropic and thyrotropic activities, but secretion of somatotropic hormone as well. Completeness of the hypophysec-

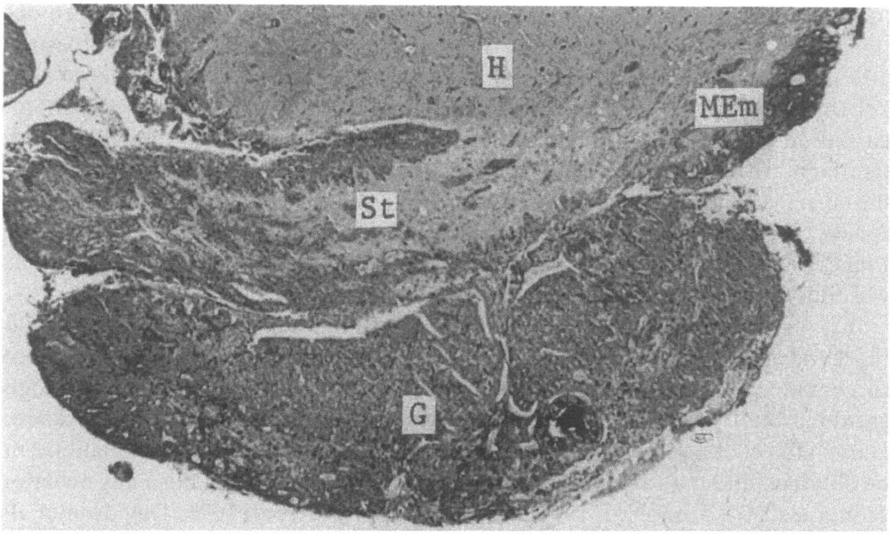

Fig. 65. Typical placement of a pars distalis graft re-transplanted under the median eminence (MEm). The graft (G) had established close contact with the MEm and the infundibular stem (St). A part of the hypothalamus (H) lies above. During most of the 3 months since the operation the animal cycled, but remained sterile. The graft is composed of healthy glandular tissue with the exception of the slender scar in the very interior. Note the epithelial cyst containing colloid darkly stained by the PAS procedure. Parasagittal section. 50X. (From Nikitovitch-Winer and Everett 1959)

tomies had been assured by the typical hypophysioprive status during the months intervening between the two operations.

Together with the return of the several trophic functions after re-transplantations in the present study, there was recovery of the characteristic inhibitory activity of the hypothalamus on prolactin secretion. None of the rats presenting regular cycles displayed spontaneous pseudopregnancy; yet when they became pregnant normal regulation of corpora lutea must be assumed.

Grafts placed under the temporal lobe (Group TL) continued to secrete prolactin. When corpora lutea were induced by injection of FSH and LH, 6 of 7 rats gave a good decidual response to trauma. With absence of gonadotropin secretion, prolactin secretion was thus renewed as the graft reorganized after the second operation.

Delayed Pseudopregnancy: An Approach to Mechanisms in Prolactin Control

The first account of a delayed pseudopregnancy response to genital stimulation in rats was that by Greep and Hisaw (1938). Electrical stimulation of the cervix during diestrus sometimes resulted in pseudopregnancies that began, not immediately, but after the following estrus. This occurred in 4 of 13 rats stimulated on D-1 and 7 of 13 rats stimulated on D-2. Much longer delays were encountered by Everett (1952, 1967) as an unexpected result of copulation during cycles in which spontaneous ovulation was blocked by pentobarbital (Fig. 66). The experiment had been designed to test for reflex ovulation. After receiving pentobarbital on the proestrus afternoon, about 50% of the females accepted coitus that night when placed with known fertile males. If further barbiturate treatment was omitted, ovulation occurred on the second night and pregnancy followed at once. On the other hand, treatment with barbiturate on the second afternoon usually prevented ovulation of the current set of follicles (Fig. 66e). There followed a short period of diestrus, then a new proestrus and estrus accompanied by ovulation of a new set of follicles. Pseudopregnancy of normal duration began at that time. The effect of the stimulus was thus "remembered" in some way for about a week.

In a very few exceptional cases, coitus induced ovulation in spite of pentobarbital blockade of the spontaneous process (Fig. 66d). Other workers have been far more successful in achieving reflex ovulation in experiments of similar design (pentobarbital blockade: Kalra and Sawyer 1970; chlorpromazine blockade: Harrington et al. 1966). In fact, it is now well documented that reflex ovulation can be obtained under a variety of circumstances in this species that normally ovulates spontaneously (Aron et al. 1961, 1966; Brown-Grant et al. 1973; Davidson et al. 1973; Leipheimer et al. 1984; Murakami et al. 1978; Smith and Davidson 1974; Taleisnik et al. 1979; Ying and Meyer 1969; Zarrow and Clark 1968). Several of the reports gave strong evidence that spontaneous ovulation and reflex ovulation involve different neural mechanisms. For example, Brown-Grant et al. (1973) noted in LLPE rats that administration of pentobarbital at 10 min or 40 min after coitus did not significantly modify the levels of circulating LH, whereas the drug promptly lowers LH when given during the normal proestrus LH surge or that induced by progesterone. A distinct quantitative influence of the degree of stimulation has been

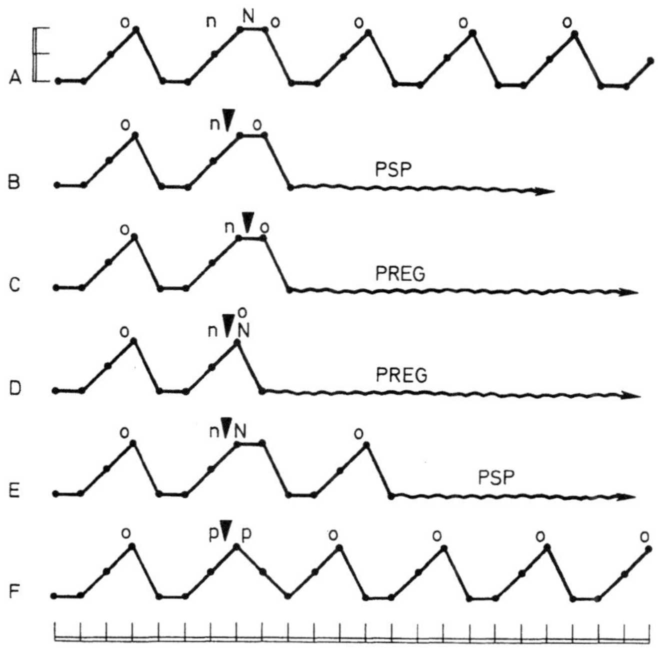

Fig. 66. Events following barbiturate block of spontaneous ovulation; influence of coitus. 24-hr intervals marked on abscissa. Key: o = ovulation; n,N = pentobarbital at critical period and heavier dose on second day; p = phenobarbital injection; black triangle = coitus. O-M rats. (From Everett 1967)

recognized (Aron et al. 1966; Brown-Grant et al. 1973; Leipheimer et al. 1984; Zarrow et al. 1968), as well as important potentiating effects of olfactory input to the hypothalamus (Aron 1979; Aron et al. 1970; Curry 1974; Yeoman 1977). One must conclude that the frequent failures of ovulation in the author's initial study (Everett 1952, 1967) resulted from inadequate stimulation, possibly resulting from insufficient numbers of copulations. Whatever the explanation, the fortuitous result allowed the first indication of the week-long mnemonic effect of the stimulus on secretion of prolactin.

Quinn and Everett (1967) extended the delayed-pseudopregnancy study by showing the delayed response after unilateral electrical stimulation of the dorsomedial-ventromedial hypothalamus (Fig. 67). A preliminary investigation (Everett and Quinn 1965) demonstrated that stimulation of the region, while unable to induce ovulation, could selectively cause pseudopregnancy when corpora lutea were induced by other means. Several cases of delayed ovulation were encountered in that work, calling for the more systematic investigation which proceeded as follows.

The rats were 4-day cyclic O-M rats at D-1, D-2, or proestrus on the day of stimulation. All stimulations were carried out under the ovulation-blocking dosage

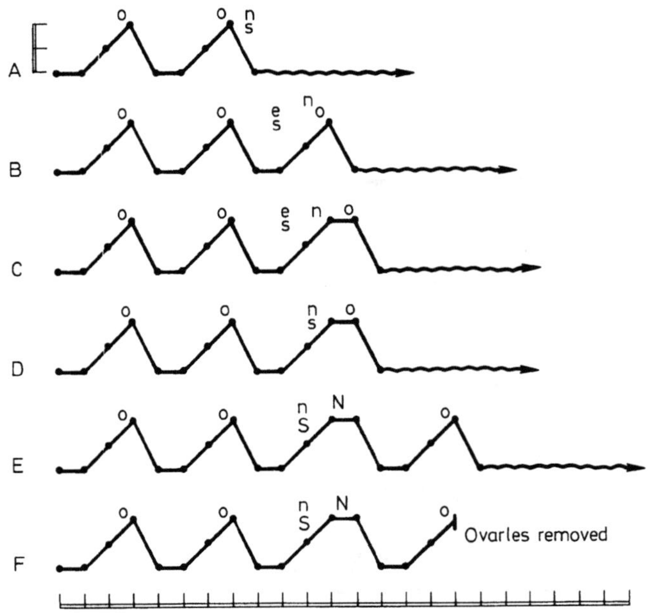

Fig. 67. Results similar to those in Fig. 66, with hypothalamic stimulation substituted for coitus. Key: n,N = pentobarbital; e = ether anesthesia; o = ovulation; s = electric stimulation for 10 min; S = same for 30 min. (Modified from Quinn and Everett 1967)

of pentobarbital, with the exception of one group of rats stimulated on D-2 under ether. In the definitive experiments (Fig. 67) the pentobarbital was injected on the proestrus afternoon just in advance of the critical period and again on the next day; the dosage on that second day was repeated after 90-min. The electrodes were the standard coaxial bipolar units of platinum-iridium, stereotaxically introduced. Matched biphasic pairs of 1-msec pulses, 100 Hz, 200 μA peak-to-peak were delivered in 30-sec trains each min. Overall duration of the stimulus varied from 10 min to 30 and 60 min. Whenever prolonged diestrus suggested pseudopregnancy, this was verified either by the decidual reaction to trauma or by the heavily mucified vaginal smear at the first estrus ending a long period of vaginal leukocytosis.

As shown in Fig. 67a, stimulation for 10 min on D-1 usually caused immediate pseudopregnancy. On the other hand, short-term delays were the rule after stimulation on D-2 (Fig. 67b and c). Short-term delay was also the rule after stimulation on proestrus under pentobarbital during the critical period (Fig. 67d). In this case both ovulation and the beginning of pseudopregnancy were delayed by the pentobarbital. In the definitive experiments, in which pentobarbital was administered on both proestrus and estrus, trials with the 10-min stimulation on proestrus were not productive; all 5 rats continued to cycle with no prolongation of diestrus (Table 37). However, long-term delays were encountered after longer stimulation; 9 of 14 rats stimulated for 30 to 60 min became pseudopregnant after completion of

Table 37. Effect of substituting electrical stimulation of the hypothalamus for the mating stimulus in the induction of delayed pseudopregnancy. (From Quinn and Everett 1967)

Experimental procedures		Results		
Proestrus	Estrus	Cycles continued	Short-term delayed pseudopregnancy	Long-term delayed pseudopregnancy
NBTL[a] + 10-min Stimulation	NBTL[b]	5	0	0
NBTL + 30-min stimulation	NBTL	2	1	5
NBTL + 60-min stimulation	NBTL	1	1	4

[a] Pentobarbital sodium on proestrus, 31.5 mg/kg, ip, before critical period.
[b] Two injections on estrus, 27 mg/kg each, before critical period and approximately 1 1/2 hr later.

the next cycle (Fig. 67e). Thus, as in the case of genital stimulation, the duration of the mnemonic effect depended upon the degree of stimulation.

The characteristic ovarian histology in a few rats that were ovariectomized at estrus in that next cycle (Fig. 67f) consistently presented absence of luteal tissue referable to the experimental cycle, a set of atretic follicles (Fig.68), and a set of fresh corpora lutea. This was likewise true in rats failing to ovulate reflexly after copulation during similar experimental cycles (Everett 1967). It should be recalled that in the absence of stimulation, from either coitus or electrical effects in the hypothalamus, pentobarbital blockade for two days is commonly followed by spontaneous ovulation during the third night (see Fig. 30, p. 54). By some mechanism presently unknown, stimuli appropriate for promoting prolactin secretion had the additional effect of causing early atresia of Graafian follicles under the conditions of these experiments. It seemed that the long delay of pseudopregnancy was in some way dependent on the absence of competent corpora lutea during the interval preceding the next cycle.

Zeilmaker (1965) devised another means for producing long-delayed pseudopregnancy. Adult female rats were each engrafted with an infantile ovary. Later on they were mated; when spermatozoa were present in the vagina the adult ovaries were removed. The infantile ovarian grafts rapidly developed, resulting 4 days later in estrus and ovulation-luteinization that initiated pseudopregnancy. As in our studies, these pseudopregnancies were of normal length, about 12–14 days.

Further investigations of delayed pseudopregnancy in this laboratory (Beach et al. 1975, 1978) were addressed to the patterns of prolactin secretion after stimulation of the cervix or the hypothalamus. Radioimmunoassay methods had made it possible to demonstrate twice-daily surges of prolactin in pseudopregnant and early pregnant rats (Butcher et al. 1972; Freeman and Neill, 1972; Freeman et al 1974). The first concern (Beach et al. 1975) was whether, in the short-delay pseudopregnancy,

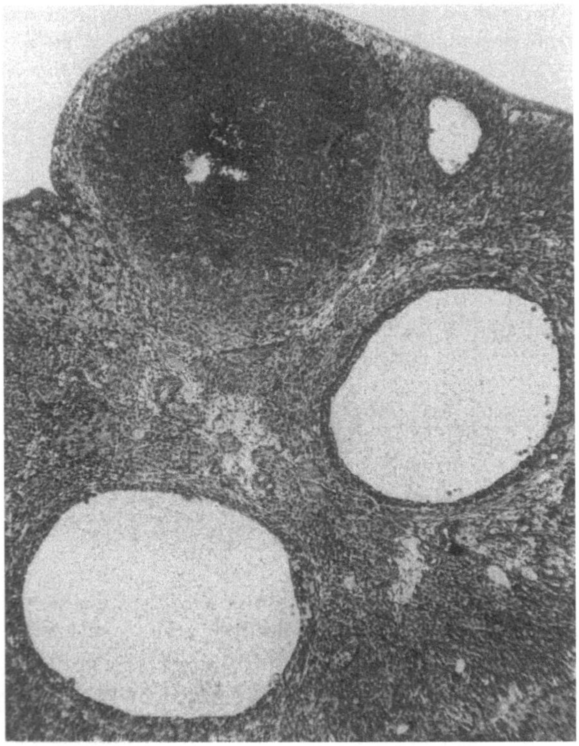

Fig. 68. Typical picture of atretic follicles in the presence of fresh corpora lutea, the presumptive start of delayed pseudopregnancy, as seen at estrus of the cycle following that in which the stimulation was administered to the DMH/VMH (Fig. 67E & F). (From Quinn and Everett 1967)

stimulation of the cervix or the hypothalamus on D-2 of the 4-day cycle is followed at once by a surge of prolactin. Corpora lutea from the preceding cycle are competent on D-2 to respond to elevated prolactin secretion (Nikitovitch-Winer and Everett 1958; Quilligan and Rothchild 1960). If genital or hypothalamic stimulation on D-2 immediately elevates prolactin secretion, why does it not produce immediate pseudopregnancy? That actually begins only when new corpora lutea have formed after the impending estrus and ovulation, as noted earlier.

The experimental animals used by Beach et al. were 4-day cyclic CD females. The cervix was electrically stimulated on D-2 at 1400h-1500h. Trunk blood samples were taken in groups of animals rapidly decapitated at 5, 15, 30, 45, and 60 min, respectively, and then at approximately 3-hour intervals through the first six days of persistent leukocyctic vaginal smears. Control data were similarly obtained from untreated rats at 3-hour intervals throughout their 4-day cycles. The results (Figs. 69 and 70) show no significant rise of serum prolactin after the cervical stimulus on D-2 and no increase until the normal rise on the proestrus afternoon (24 h after the stimulation). A significant nocturnal surge occurred spontaneously early on estrus

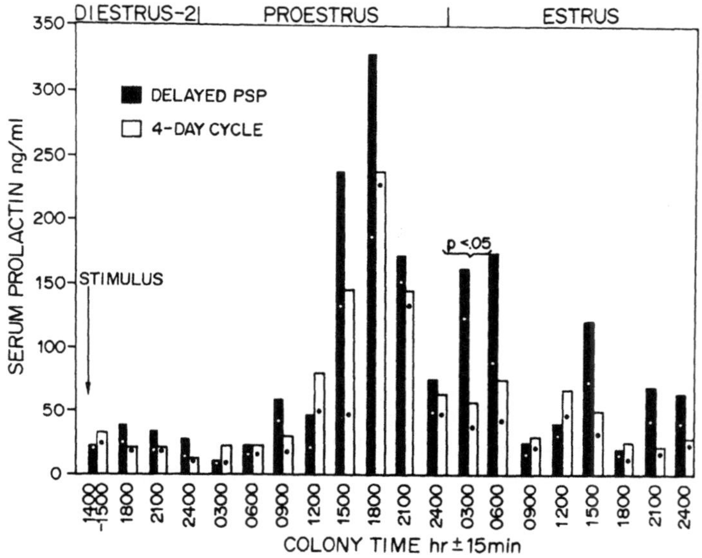

Fig. 69. Serum prolactin concentrations in short-delay pseudopregnancy following cervical stimulation at 1400h on diestrus-2 through the "interim" of proestrus and estrus, closely similar to values observed at comparable times in normal 4-day cyclic control rats. The 1400–1500h group includes all samples collected within that time range. CD rats. (From Beach et al. 1975)

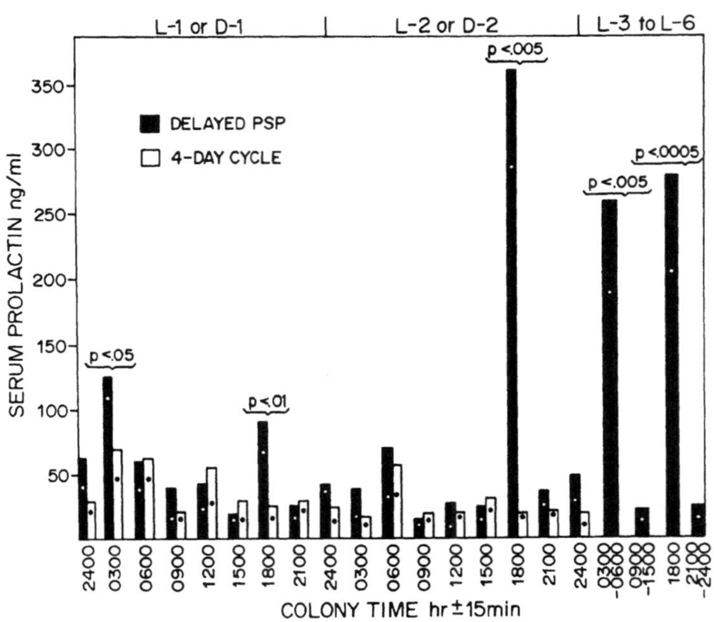

Fig. 70. Continues from Fig. 69 after the "interim" of proestrus and estrus and throughout the first 6 days of leukocytic vaginal smears of pseudopregnancy (L-1 to L-6). Prolactin values on L-1 and L-2 are compared with values on D-1 and D-2 in cyclic control rats. The pooled peak values are compared statistically with values at nadir. Dots in bars mark the SE. (From Beach et al. 1975)

and small surges appeared the next day at both 0300h and 1800h. Large diurnal and nocturnal surges occurred each day thereafter. Other rats were cervically stimulated on the afternoon of estrus, blood samples being taken as before at 3-hour intervals. Again there was no immediate increase of prolactin above control values, the first significant rise occurring at midnight. Regular twice-daily surges were evident from the second day of diestrus onward.

The second concern (Beach et al. 1978) was whether in the case of long-term delay there are nocturnal and diurnal surges of prolactin during the interval between stimulation and the succeeding proestrus, in spite of the absence of corpora lutea. To test this, hypothalamic stimulation was the method of choice. In adult CD females pentobarbital was administered at the critical hours on proestrus and on the next two days to ensure the absence of luteinization in the current cycle. On the third afternoon, the experimental rats were stimulated through the platinum coaxial electrode aimed unilaterally at the dorsomedial-ventromedial hypothalamus (Figs. 71 and 72). Parameters of stimulation were those used by Everett and Quinn (1966) and Quinn and Everett (1967), with overall duration standardized at 30 min. Control rats were prepared in the same way; the electrode was left in place for 30 min without passage of current. Since pilot studies showed that only ~60% of animals stimulated on the third day of pentobarbital treatment would become pseudopregnant, blood sampling in individuals was sequential by means of intracardiac cannulas implanted before stimulation. Beginning at 0300h on the early morning after stereotaxis, samples were taken at the times indicated in Fig. 73. In the stimulated rats that later

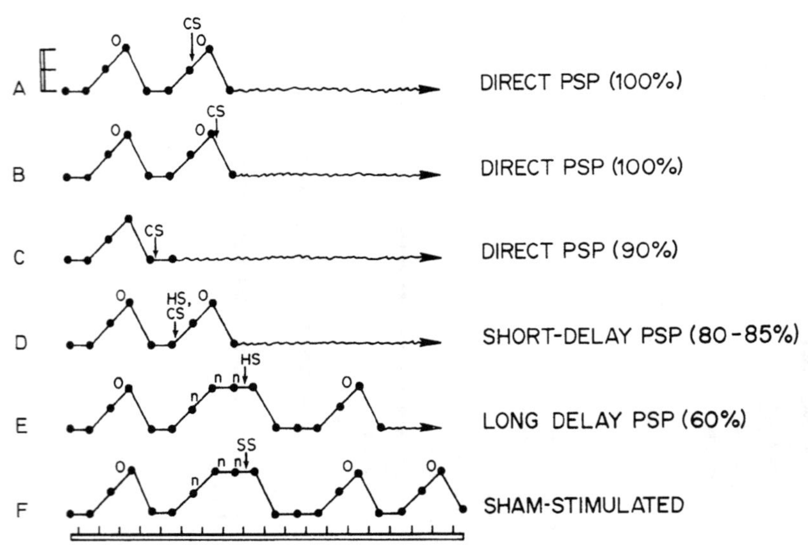

Fig. 71. Direct or delayed pseudopregnancy after stimulation of the cervix or the hypothalamus (VMH/DMH). Units on the abscissa mark 24h intervals. Key: CS = cervical stimulation; HS = hypothalamic stimulation; n = ovulation blockade with pentobarbital; o = ovulation; SS = sham stimulation. CD rats. (From Beach et al. 1978)

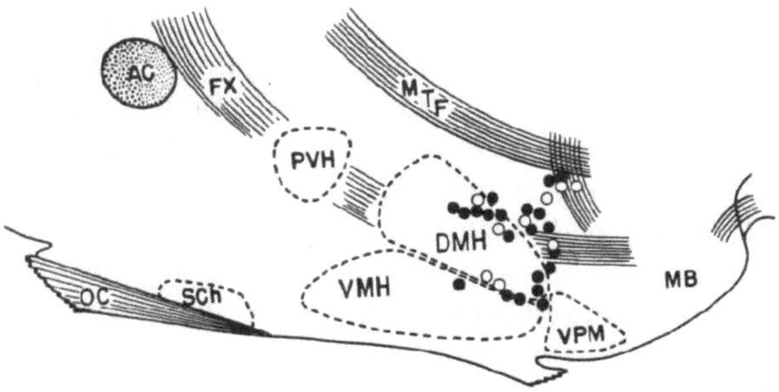

Fig. 72. Hypothalamic electrode sites in animals stimulated on the third day of ovulation blockade. See Fig. 71. Key: o = stimulus producing long delay of PSP; ● = stimulated animals continuing to cycle. (From Beach et al. 1978)

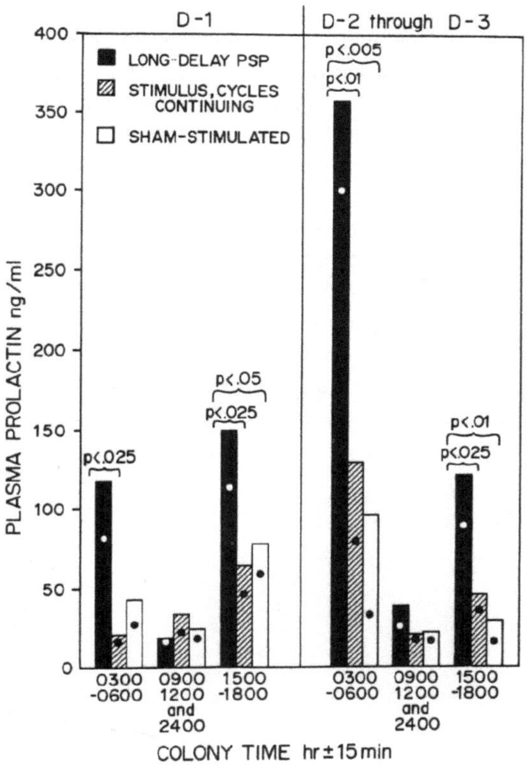

Fig. 73. PRL secretion during the first 3 days after hypothalamic stimulation on the third day of ovulation blockade. PRL values in rats later manifesting long-delay PSP are compared with values in stimulated or sham-stimulated rats that continued to cycle. Each bar represents at least 5 blood samples from indwelling cardiac catheters. (From Beach et al. 1978)

became pseudopregnant there were twice-daily surges of prolactin that were significantly higher than in sham stimulated rats or those continuing to cycle.

Experiments by de Greef and Zeilmaker (1976) show that prolactin surges occurring during the long delay interval do not constitute an essential prelude to the forthcoming pseudopregnancy. Treatment with ergocornine at the beginning of normal pseudopregnancy suppressed both nocturnal and diurnal surges for two days, yet after renewed proestrus both surges appeared in the 60% of the rats which then became pseudopregnant. No surges appeared in rats that continued to cycle. In other experiments, the removal of *in situ* ovaries from animals bearing ovarian grafts or the removal of corpora lutea in early pseudopregnancy resulted in nocturnal but not diurnal surges during the delay period. The lack of diurnal surges may have reflected inadequate levels of ovarian steroids, whereas in the experiments of Beach et al. (1978) the ovaries remained undamaged. The diurnal surges are known to be steroid dependent (Freeman et al. 1974).

Whether or not the prolactin surges occur during the delay period, the more important consideration is that the effect of the stimulus is stored for about a week, presumably in the central nervous system. Zeilmaker (1965) indicated an upper time limit of 7–9 days, as shown in rats bearing ovarian grafts. After infertile copulation and removal of the *in situ* ovaries at progressively later times thereafter, pseudopregnancies occurred only when that operation was performed on days 0, 1, or 2, not on days 3 or 4. Equally important is the quantitative relationship between the degree of stimulation and the length of time that the stimulus is "remembered". Especially significant is the fact that this relationship appeared after both peripheral (i.e. genital) and hypothalamic stimulation.

Overall, the studies of delayed pseudopregnancy in this laboratory and elsewhere have clearly demonstrated that the induction of pseudopregnancy does not constitute a "reflex" in the usual sense. There is not even an immediate release of prolactin. Rather, the process involves some protracted change in central neural physiology, as if by a change in neurochemistry. Whatever that change may be, it promotes the regular nocturnal prolactin surges and permits the hypothalamus to produce the diurnal surges when influenced by ovarian steroids.

Summary

With respect to the ovarian status and the estrous cycle as expressed by the vaginal cytology, it was emphasized that there is only a rough correspondence between the *vaginal stages* and the *days* of the cycle. The 4-day cycle tends to be the most frequent, with the 5-day cycle a normal variant. Spontaneous persistent vaginal estrus (SPE), an acyclic state occurring in older adult rats, reflects the presence in the ovaries of large vesicular follicles failing to luteinize and, hence, the absence of corpora lutea once the condition becomes well established. Pseudocyclic fluctuations in the vaginal smear during SPE resemble those in ovariectomized rats treated chronically with estrogen.

The age of onset of SPE varies among different rat strains, appearing as early as 5 or 6 months in the DA and CD strains, but rarely before 12 months in the O-M strain. O-M/DA hybrids were intermediate. In the DA strain there was also a marked influence of length of daily illumination on the occurrence of SPE: exposure to 10 hours or less of light per day restored cycling to rats that had already shown SPE while exposed to longer days.

An hereditary influence of age was also apparent in the rapidity with which continuous illumination induced persistent estrus (LLPE). Young DA females rapidly entered LLPE within 10 days, while young O-M females remained cyclic for 5 weeks. Hybrids again were intermediate. At middle age, O-M rats became as responsive as young rats of the DA strain. Like SPE in DA rats, LLPE was reversible, for estrous cycles returned after daily exposure to light was reduced. The special sensitivity of DA rats to lighting may have been a trait acquired from outcrossing with wild gray rats in years past.

Normal cycling could be restored in SPE rats by daily injection of progesterone at low dosage. The same effect followed isolated treatments with progesterone upon return of proestrus/estrus after interruption of SPE. This was the first demonstration of positive feed-back of progesterone, the first sign of its biphasic action, and an indication that progesterone facilitates the action of estrogen in promoting ovulation. When progesterone treatment was delayed after proestrus/estrus there was progressively lower effectiveness during the next 10 days.

Indirect support of regular ovulatory cycles resulted from treating DA SPE rats with prolactin (PRL) daily at low doses, provided that an initial set of corpora lutea was first induced by other means. A similar effect of *intrinsic* PRL was obtained in middle-aged *CD* SPEC rats; the mere introduction of a set of corpora lutea by injecting LH resulted typically in a sequence of several ovulatory cycles before the eventual return of SPE. Cycling could be prevented by treating with bromocryptine and that could be overcome by adding extrinsic PRL at the same time. Thus it seems that low-grade secretion of PRL, through its luteotropic action and the low-level production of progesterone, tends to protect against SPE.

Comparative studies of corpora lutea in cyclic DA rats and O-M rats showed that DA corpora lutea were comparatively lacking in cholesterol content. The postovulatory DA ovaries lacked the marked accumulation of cholesterol and the related fatty degenerative areas in the next older corpora lutea as seen in normal (O-M) rats. The same lack was evident in DA SPE ovaries after progesterone-induced cycles. However, after PRL-sustained cycles the DA corpora lutea resembled those of O-M rats. Subsequent findings indicated that luteal storage of cholesterol depends on both PRL and LH.

In pregnant or pseudopregnant O-M rats, EB given on day 4 or day 5 induced ovulation during the second night, accompanied by marked storage of cholesterol in the primary corpora lutea. Hypophysectomy on the day after the estrogen administration prevented both effects, but that could be counteracted by injecting LH at the time of the operation. Cholesterol accumulation in response to an LH injection could be overcome by giving an excess of PRL at the same time. Evidently the luteal storage of cholesterol depends on the LH/PRL ratio, with LH in the role of increasing the store and PRL governing the rate of utilization of the sterol in steroid synthesis. Significantly the cholesterol deposits induced by LH disappeared within 24 hours when pregnancy continued.

In 5-day cyclic O-M rats ovulation could be advanced about 24 hours by injecting progesterone on diestrus day 3 (D-3) or by injecting estradiol benzoate (EB) or implanting pellets of estradiol-17β on D-2. On the contrary, this effect of estrogen was not usual in young DA females or DA SPE rats during cycles induced with progesterone. But if DA SPE rats received 1.5 mg progesterone daily to simulate pseudopregnancy, EB injection induced ovulation and/or luteinization within 48 hours.

The fact that both SPE and LLPE in DA rats were reversible by modifying the daily exposure to light suggested that the central nervous system participates in the induction of ovulation. This became more apparent with the demonstration that drugs which block the ovulation reflex in rabbits, such as Dibenamine and atropine, will block both spontaneous and steroid-induced ovulation in rats. A 'critical period' during the proestrus afternoon was defined by use of atropine and pentobarbital, which blocked ovulation consistently if given before 1400h. After pentobarbital blockade on proestrus a new critical period occurred next afternoon; blockade on both days delayed ovulation to the third night, while blockade for three days prevented ovulation throughout the current cycle and a new proestrus typically appeared 2 days later on. A potential afternoon critical period on D-3 of the 5-day cycle, approximately 24 hours early was shown by atropine blockade of steroid induced ovulation. This, together with the day-to-day experience with pentobarbital, was first evidence of the circadian periodicity of neural apparatus controlling phasic release of LH in this species.

Atropine injection or hypophysectomy at different times after 1400h during the proestrus afternoon showed that the effects of both treatments were identical. As the time of atropine injection or hypophysectomy progressed, greater numbers of rats ovulated, showing that different rats began the LH surge at different times. The fact that roughly 20% of the animals showed partial effects indicated that the least amount of LH needed for full ovulation (the ovulation quota) is released in ~30 min. Furthermore, the amount released was evidently proportional to the duration of the atropine-sensitive stimulus. The atropine-sensitive stimulus and LH release proceeded approximately in parallel; this stimulus in rats thus differs from the brief trigger indicated for the rabbit ovulation reflex.

Circadian periodicity of the phasic LH surge in rats, its blockage by barbiturates and other drugs having central nervous activity, the dependence of its timing on environmental lighting, all pointed to its control by the brain. Direct evidence came from demonstration that the natural ovulatory surge can be reproduced by brain stimulation in rats whose spontaneous LH surges are absent, as in late diestrus, pseudopregnancy, SPE, LLPE, TPPE, or because of pharmacologic blockade. Stimulations were carried out with either electrochemical (ECS) or electrical (ELS) methods. Mapping studies with ECS indicated a funnel-shaped septal-preoptic-tuberal system (SPTS), rostrally diffuse and caudally convergent on the medial basal tuber. The degree of involvement of this system apparently determines the amount of LH released in a given time. Massive bilateral ECS foci in the MPOA could release the ovulatory quota of LH before hypophysectomy at 30 min. Smaller unilateral lesions released the quota within 45 min in proestrous rats whereas in D-3 rats an additional 15–20 min was needed. (Dose-response studies showed that D-3 rats require more LH). The relative effectiveness of ECS in the MPOA and the MBT could not be evaluated because of the fact that non-specific damage to the MBT by knife cuts or radiofrequency lesions can induce an LH surge. Comparison of the two regions was feasible with electrical stimulation, however.

Electrical stimulation held advantage over ECS, not only for the relative lack of destruction, but also for allowing control over the stimulus duration. A disadvantage was suggested by the consistently lower effectiveness of electrical stimulation in the MPOA than in the ARC/ME. This was interpreted to mean that a stimulus of given intensity can reach a relatively greater proportion of the SPTS where it converges to the ARC/ME. There were positive dose-response relationships as measured by the amounts of LH released and by the numbers of rats ovulating. The curvilinear patterns of circulating LH accompanying stimulation resembled the normal proestrous surge; the amounts released at 45–60 minutes corresponded to the ovulation "quota", while amounts released during the second hour of stimulation resembled the normal excess. The parallelism of increments of LH during MPOA and ARC/ME stimulations implied that the entire SPTS serves as a functional unit.

In SPE rats, ovulation could readily be induced by either ECS or ELS. An interesting difference between SPE and proestrous subjects was that during the second hour of ELS stimulation LH levels in SPE rats tended to plateau while they continued to rise in proestrous rats. It is especially significant that ovulation was readily induced by brain stimulation whereas acute interruption of SPE by progesterone was only rarely accompanied by ovulation. Therefore, during SPE the AP has an adequate content of LH and there must be an adequate supply of LHRH awaiting activation of the SPTS.

Similar conclusions are called for by the marked similarity of LH levels after MPOA stimulation under blockade of spontaneous ovulation by pentobarbital, morphine, chlorpromazine or atropine. None of these drugs has its primary blocking action on the SPTS, on the availability of LHRH or its release, on the capacity of the AP to release LH, or on the responsiveness of the ovaries.

Studies of pseudopregnancy and the controls of PRL secretion involved effects of transplantation of the AP away from hypothalamic influence primarily to the renal capsule, effects of damage to the AP *in situ* or to the hypothalamus, effects of re-transplantation of the established graft from kidney to the renewed influence of the hypothalamus, and the phenomenon of delayed pseudopregnancy. AP transplants to the kidney continued to maintain corpus luteum function for weeks or months while trophic secretions other than PRL were greatly diminished or lost. This was first evidence of hypothalamic inhibition of PRL secretion.

Simple exposure of the AP by the parapharyngeal approach failed to induce pseudopregnancy unless there was some slight damage to the gland. Stalk section induced pseudopregnancy, but cycling returned unless a barrier was inserted to prevent regeneration of the portal vessels. Accidental damage to the MBT could also cause temporary pseudopregnancy.

Re-transplantation of established AP grafts from the kidney to the median eminence and renewed blood supply from the portal circulation resulted in reappearance of gonadotropic, TSH, and ACTH secretions as PRL secretion diminished. Estrous cycles were restored and many of the rats became pregnant. The restoration occurred in spite of the dual insult to the gland accompanied each time by massive infarction; healthy parenchyma regenerated from the thin shell that remained. This experiment by Nikitovitch-Winer re-emphasized the special importance of the portal blood supply for normal pituitary functions.

Although briefly delayed pseudopregnancy responses had been noted by others as the result of cervical stimulation in late diestrus, much longer delays were encountered in mating experiments with proestrous rats during pentobarbital blockade. Although a few rats ovulated reflexly in response to copulation, in most cases ovulation failed to occur in the current cycle and pseudopregnancy began about a week later after spontaneous ovulation in a new cycle. Similar results were obtained by electrical stimulation of the hypothalamus in sites not conducive to ovulation. Short-term delays were produced by 10-min stimulation, long-term delays by 30- to 60-min stimulation. This quantitative difference is like that encountered with genital stimulation, in which artificial cervical stimulation was inadequate for producing the long-term delay that could be obtained with copulation.

Prolactin secretion was not immediately induced by cervical stimulation or by stimulation of the hypothalamus. Instead there were twice-daily surges of PRL, like those of normal pseudopregnancy (in spite of the absence of recently formed corpora lutea during the long-delay interval). Thus the induction of pseudopregnancy does not result from rapid reflex induction of PRL secretion but from some protracted change in CNS physiology as if by a shift in neurochemistry.

References

Aisen P (1977) Some physiological aspects of iron metabolism. Ciba Found Symp (New Series) 51:1–14

Aiyer MS, Chiappa SA, Fink G (1974) A priming effect of luteinizing hormone releasing factor on the anterior pituitary gland in the female rat. Endocrinology 62:573–588

Allen E (1922) The oestrous cycle in the mouse. Amer J Anat 30:297–371

Arai Y (1971) A possible process of the secondary sterilization: delayed anovulation syndrome. Experientia 27:463–464

Aron CL (1979) Mechanisms of control of the reproductive function by olfactory stimuli in female mammals. Physiol Rev 59:229–284

Aron CL, Asch G, Asch L (1961) Déclenchement de la ponte ovulaire et de la lutéinisation par le rapprochement sexual chez les Mammiferès dits "à ponte spontanée". Expériences chez la Ratte. Compt rend Soc Biol 155:2173–2176

Aron CL, Asch G, Roos J (1966) Triggering of ovulation by coitus in the rat. Internat Rev Cytol 20:139–172

Aron CL, Roos J, Asch G (1970) Effect of removal of olfactory bulbs on mating behaviour and ovulation in the rat. Neuroendocrinology 6:109–117

Aschheim P (1961) La pseudogestation à répétition chez les Rattes séniles. Compt rend Acad Sci 253:1988–1990

Aschheim P (1965) La réactivation de l'ovaire des rattes séniles en oestrus permanent au moyen d'hormones gonadotropes ou de la mise à l'obscurité. Compt rend Acad Sci 260:5627–5630

Asdell SA, Crowell MF (1935) The effect of retarded growth upon the sexual development of rats. J Nutrition 10:13–24

Astwood EB (1941) The regulation of corpus luteum function by hypophysial gonadotrophin. Endocrinology 28:309–320

Baldwin DM, Sawyer CH (1979) Light synchronization of the preovulatory LH surge in adrenalectomized rats. Proc Soc Exp Biol Med 161:295–298

Banks JA, Freeman ME (1978) The temporal requirement of progesterone on proestrus for extinction of the estrogen-induced daily signal controlling luteinizing hormone release in the rat. Endocrinology 102:426–432

Barr GD, Barraclough CA (1978) Temporal changes in medial basal hypothalamic LH-RH correlated with plasma LH during the rat estrous cycle and following electrochemical stimulation of the medial preoptic area in pentobarbital-treated proestrous rats. Brain Res 148:413–423

Barraclough CS, Gorski RA (1961) Evidence that the hypothalamus is responsible for androgen-induced sterility in the female rat. Endocrinology 68:68–79

Barraclough CA and Sawyer CH (1955) Inhibition of the release of pituitary ovulatory hormone in the rat by morphine. Endocrinology 57:329–337

Barraclough CA and Sawyer CH (1957) Blockade of the release of pituitary ovulating hormone in the rat by chlorpromazine and reserpine: possible mechanisms of action. Endocrinology 61:341–351

Barraclough CA, Yrarrazawal A, Hatton R (1964) A possible hypothalamic site of action of progesterone in the facilitation of ovulation in the rat. Endocrinology 75:838–845

Beach JE, Tyrey L, Everett JW (1975) Serum prolactin and LH in early phases of delayed versus direct pseudopregnancy in the rat. Endocrinology 96:1241–1246

Beach JE, Tyrey L, Everett JW (1978) Prolactin secretion preceding delayed pseudopregnancy in rats after electrical stimulation of the hypothalamus. Endocrinology 103:2247–2251

Blake CA (1974) Differentiation between the 'critical period,' the 'activation period' and the 'potential activation period' for neurohumoral stimulation of LH release in proestrous rats. Endocrinology 95:572–578

Blake CA (1976a) A detailed characterization of the proestrous luteinizing hormone surge. Endocrinology 98:445–450

Blake CA (1976b) Simulation of the proestrous luteinizing hormone (LH) surge after infusion of LH-releasing hormone in phenobarbital blocked rats. Endocrinology 98:451–460

Blake CA, Sawyer CH (1972) Ovulation blocking actions of urethane in the rat. Endocrinology 91:87–94

Bloch S, Flury E (1959) Untersuchungen über Klimakterium und Menopause an Albino-Ratten. Gynaecologia 147:414–438

Boehm N, Plas-Roser S, Aron CL (1984) Does corpus luteum function autonomously during estrous cycle in the rat? A possible involvement of LH and prolactin. J Steroid Biochem 20:663–670

Boling JL, Blandau RJ, Soderwall AL, Young WC (1941) Growth of the Graafian follicle and the time of ovulation in the albino rat. Anat Rec 79:313–331

Bridges RD, Goldman BD (1975) Diurnal rhythms in gonadotropins and progesterone in lactating and photoperiod induced acyclic hamsters. Biol Reprod 13:617–622

Browman LG (1937) Light in its relation to activity and oestrous rhythms in the albino rat. J Exp Zool 75:375–388

Brown-Grant K (1967) The effects of a single injection of progesterone on the oestrous cycle, thyroid gland activity and uterus-plasma concentration ratio for radio-iodide in the rat. J Physiol, London 190:101–121

Brown-Grant K (1969a) The induction of ovulation during pregnancy in the rat. J Endocrinol 43:529–538

Brown-Grant K (1969b) The induction of ovulation by ovarian steroids in the adult rat. J Endocrinol 43:553–562

Brown-Grant K, Davidson JM, Grieg F (1973) Induced ovulation in albino rats exposed to constant light. J Endocrinol 57:7–22

Bunn JP, Everett JW (1957) Ovulation in persistent-estrous rats after electrical stimulation of the brain. Proc Soc Exp Biol Med 96:369–371

Burrows H (1949) Biological Actions of Sex Hormones. Cambridge Univ Press, Cambridge, UK

Butcher RL, Fugo NW, Collins WE (1972) Semicircadian rhythm in plasma levels of prolactin during early gestation in the rat. Endocrinology 90:1125–1127

Butcher RL, Collins WE, Fugo NW (1975) Altered secretion of gonadotropins and steroids resulting from delayed ovulation in the rat. Endocrinology 96:576–586

Caligaris L, Astrada JJ, Taleisnik S (1971a) Release of luteinizing hormone induced by estrogen injection into ovariectomized rats. Endocrinology 88:810–815

Caligaris L, Astrada JJ, Taleisnik S (1971b) Biphasic effect of progesterone on the release of gonadotropin in rats. Endocrinology 89:331–337

Clemens JA, Amenomori Y, Jenkins K, Meites J (1969) Effects of hypothalamic stimulation, hormones, and drugs on ovarian function in old female rats. Proc Soc Exper Biol Med 132:561–563

Clemens JA, Smalstig EB, Sawyer BD (1976) Studies on the role of the preoptic area in the control of reproductive function in the rat. Endocrinology 99:728–735

Colombo JA, Saporta S (1980) Increased local uptake of 2-deoxyglucose after electrochemical or direct deposition of iron into the rat brain. Exper Neurol 70:427–437

Colombo JA, Whitmoyer DI, Sawyer CH (1974) Local changes in multiple unit activity induced by electrochemical means in preoptic and hypothalamic areas in the female rat. Brain Res 71:1175–1183

Colombo JA, Whitmoyer DI, Sawyer CH (1975) Local effects of iron deposition on multiple unit activity in the female rat brain. Brain Res 96:88–92

Conrad LCA, Pfaff DW (1976) Efferents from medial basal forebrain and hypothalamus in the rat. I. An autoradiographic study of the medial preoptic area. J Comp Neurol 169:185–220

Cordova T, Ayalon D, Lander N, Mechovlam R, Nir I, Puder M and Lindner HR (1980) The ovulation blocking effect of cannabinoids: structure-activity relationships. Psychoneuroendocrinol 5:53–62

Critchlow BV (1957) Ovulation induced by hypothalamic stimulation in the rat. Anat Rec 127:283 (Abstract)

Critchlow V (1958) Ovulation induced by hypothalamic stimulation in the anesthetized rat. Amer J Physiol 195:171–174

Curry JJ (1974) Alterations in incidence of mating and copulation-induced ovulation after olfactory bulb ablation in female rats. J Endocrinol 62:245–250

Davidson JM, Smith ER, Bowers CY (1973) Effects of mating on gonadotropin release in the female rat. Endocrinology 93:1185–1192

de Greef WJ, van der Schoot P (1979) Examination of the role of 'non-functional' corpora lutea in the female rat. J Endocrinol 83:205–209

de Greef WJ, Zeilmaker GH (1976) Prolactin and delayed pseudopregnancy in the rat. Endocrinology 98:305–310

del Castillo EB, Calatroni CJ, (1930) Cycle sexual périodique et folliculine. Compt rend Soc Biol 104:1024–1028

del Castillo EB, di Paola G (1942) Cyclical vaginal response to the daily administration of estradiol in castrated rats. Endocrinology 30:48–53

Dempsey EW, Searles HF (1943) Environmental modification of certain endocrine phenomena. Endocrinology 32:119–128

Desclin L (1950) A propos du méchanisme d'action des oestrogènes sur le lobe antérieur de l'hypophyse chez le rat. Ann Endocrinol 11:656–659

Dey FL (1941) Changes in ovaries and uteri in guinea pigs with hypothalamic lesions. Amer J Anat 69:61–87

Dey FL (1943) Evidence of hypothalamic control of hypophyseal gonadotropin function in the female guinea pig. Endocrinology 33:75–82

Dey FL, Fisher C, Berry CM, Ranson SW (1940) Disturbances in reproductive functions caused by hypothalamic lesions in female guinea-pigs. Amer J Physiol 129:39–46

Dohler KD, Wuttke W (1974) Total blockade of phasic pituitary prolactin release in rats: effect on serum LH and progesterone during the estrous cycle and pregnancy. Endocrinology 27:681–686

Dyer RG, Burnet F (1976) Effects of ferrous ions on preoptic area neurons and luteinizing hormone secretion in rat. J Endocrinol 69:247–254

Dyer RG, Mayes LC (1978) Electrical stimulation of the hypothalamus: new observations on the parameters necessary for ovulation in rats anaesthetized with pentobarbitone during the pro-oestrous 'critical period'. Exp Brain Res 33:583–592

Engle ET (1931) The pituitary gonadal relationship and the problem of precocious sexual maturity. Endocrinology 15:405–420

Eskay RL, Mical RS, Porter JC (1977) Relationship between luteinizing hormone releasing hormone concentration in hypophysial portal blood and luteinizing hormone release in intact, castrated and electrochemically-stimulated rats. Endocrinology 100:263–270
Evans HM, Long JA (1921) Effect of anterior lobe of hypophysis administered intraperitoneally upon growth, maturity, and oestrous cycles in the rat. Anat Rec 21:61 (Abstract)
Evans HM, Simpson ME, Lyons WR, Turpeinen K (1941) Anterior pituitary hormones which favor production of traumatic uterine placentoma. Endocrinology 28:933–945
Everett JW (1939) Spontaneous persistent estrus in a strain of albino rats. Endocrinology 25:123–127
Everett JW (1940) The restoration of ovulatory cycles and corpus luteum formation in persistent-estrous rats by progesterone. Endocrinology 27:681–686
Everett JW (1942) Certain functional interrelationships between spontaneous persistent estrus, "light estrus" and short-day anestrus in the albino rat. Anat. Rec. 82:403 (Abstract)
Everett JW (1943) Further studies on the relationship of progesterone to ovulation and luteinization in the persistent-estrous rat. Endocrinology 32:285–292
Everett JW (1944a) Evidence in the normal albino rat that progesterone facilitates ovulation and corpus-luteum formation. Endocrinology 34:136–137
Everett JW (1944b) Evidence suggesting a role of the lactogenic hormone in the estrous cycle of the albino rat. Endocrinology 35:507–520
Everett JW (1945) The microscopically demonstrated lipids of cyclic corpora lutea in the rat. Amer J Anat 77:293–323
Everett JW (1947) Hormonal factors responsible for deposition of cholesterol in the corpus luteum of the rat. Endocrinology 41:364–377
Everett JW (1948) Progesterone and estrogen in the experimental control of ovulation time and other features of the estrous cycle in the rat. Endocrinology 43:389–405
Everett JW (1951) Effects of estrogen-progesterone synergy on thresholds and timing of the "LH-release apparatus" of the female rat. Anat Rec 109:291 (Abstract)
Everett JW (1952) Presumptive hypothalamic control of spontaneous ovulation. Ciba Found. Colloquia on Endocrinology 4:167–178
Everett JW (1954) Luteotrophic function of autografts of the rat hypophysis. Endocrinology 54:685–690
Everett JW (1956a) Functional corpora lutea maintained for months by autografts of rat hypophyses. Endocrinology 58:786–796
Everett JW (1956b) The time of release of ovulating hormone from the rat hypophysis. Endocrinology 59:580–585
Everett JW (1961) Preoptic region of the brain and its relation to ovulation. In: Control of Ovulation (CA Vilke, ed) Pergamon Press, Oxford, UK, pp 101–112
Everett JW (1964) Preoptic stimulative lesions and ovulation in the rat: 'thresholds' and LH-release time in late diestrus and proestrus. In: Major Problems in Neuroendocrinology (E Bajusz & G Jasmin, eds) S Karger, Basel/New York, pp 346–366
Everett JW (1965) Ovulation in rats from preoptic stimulation through platinum electrodes. Importance of duration and spread of stimulus. Endocrinology 76:1195–1201
Everett JW (1967) Provoked ovulation or long-delayed pseudopregnancy from coital stimuli in barbiturate-blocked rats. Endocrinology 80:145–154
Everett JW (1969) Diencephalic regulation of corpus luteum formation and secretory activity. In: Ovum Implantation (MC Shelesnyak and GJ Marcus, eds) pp 213–131. Gordon and Breach, New York
Everett JW (1970) Photoregulation of the ovarian cycle in the rat. In: La Photorégulation chez les Oiseaux et les Mammifères (J Benoit and I Assenmacher, eds) Centre National de la Recherche Scientifique, Paris, p 387

Everett JW (1980) Reinstatement of estrous cycles in middle-aged spontaneously persistent-estrous rats: Importance of circulating prolactin and the resulting facilitative action of progesterone. Endocrinology 106:1691–1696

Everett JW (1984) Further study of oestrous cycles that follow interruption of spontaneous persistent oestrus in middle-aged rats. J Endocrinol 102:271–276

Everett JW, Krey LC, Tyrey L (1973) The quantitative relationship between electrochemical preoptic stimulation and LH release in proestrous *versus* late diestrous rats. Endocrinology 93:947–953

Everett JW, Nichols DC (1968) The timing of ovulatory release of gonadotropin induced by estrogen in pseudopregnant and diestrous cyclic rats. Anat Rec 160:346 (Abstract)

Everett JW, Quinn DL (1966) Differential hypothalamic mechanisms inciting ovulation and pseudopregnancy in the rat. Endocrinology 78:141–150

Everett JW, Quinn DL, Tyrey L (1976) Comparative effectiveness of preoptic and tuberal stimulation for luteinizing hormone release and ovulation in two strains of rats. Endocrinology 98:1302–1308

Everett JW, Radford HM (1961a) Chemical versus electrical excitation of the preoptic hypothalamus in experimental induction of ovulation in the rat. Anat Rec 139:226 (Abstract)

Everett JW, Radford HM (1961b) Irritative deposits from stainless steel electrodes in the preoptic rat brain causing release of pituitary gonadotropin. Proc Soc Exper Biol Med 108:604–609

Everett JW, Radford HM, Holsinger J (1964) Electrolytic irritative lesions in the hypothalamus and other forebrain areas: effects on luteinizing hormone release and the ovarian cycle. In: Hormonal Steroids, Biochemistry, Pharmacology and Therapeutics: Proceedings of the First International Congress on Hormonal Steroids (L Martini and A Pecile, eds), Vol 1: pp 235–246. Academic Press, New York

Everett JW, Sawyer CH (1949) A neural timing factor in the mechanism by which progesterone advances ovulation in the cyclic rat. Endocrinology 45:581–595

Everett JW, Saywer CH (1950) A 24-hour periodicity in the "LH-release apparatus" of female rats, disclosed by barbiturate sedation. Endocrinology 47:198–218

Everett JW, Sawyer CH (1953) Estimated duration of the spontaneous activation which causes release of ovulating hormone from the rat hypophysis. Endocrinology 52:83–92

Everett JW, Sawyer CH, Markee JE (1949) A neurogenic timing factor in control of the ovulatory discharge of luteinizing hormone in the cyclic rat. Endocrinology 44:234–250

Everett JW, Tejasen T (1967) Time factor in ovulation blockade in rats under differing lighting conditions. Endocrinology 80:790–792

Everett JW, Tyrey L (1977) Induction of LH release and ovulation in rats by radiofrequency lesions of the medial basal tuber cinereum. Anat Rec 187:575 (Abstract)

Everett JW, Tyrey L (1981) Comparative increments of circulating luteinizing hormone in rats with increasing duration of electrical stimulation in medial preoptic or medial basal tuberal sites. Endocrinology 109:691–696

Everett JW, Tyrey L (1982a) Comparison of luteinizing hormone surge responses to ovarian steroids in cyclic and spontaneously persistent estrous rats of middle age. Biol Reprod 26:663–672

Everett JW, Tyrey L (1982b) Similarity of luteinizing hormone surges induced by medial preoptic stimulation in female rats blocked with pentobarbital, morphine, chlorpromazine, or atropine. Endocrinology 111:1979–1985

Everett JW, Tyrey L (1983) Comparable surges of luteinizing hormone induced by preoptic or medial basal tuberal electrical stimulation in spontaneously persistent estrous or cyclic proestrous rats. Endocrinology 112:2015–2020

Fee AR, Parkes AS (1929) Studies on ovulation. I. The relation of the anterior pituitary body to ovulation in the rabbit. J Physiol (London) 67:383–388

Fevold HL, Hisaw FL, Greep RO (1936) Effect of oestrin on the activity of the anterior lobe of the pituitary. Amer J Physiol 114:508–513

Fink G, Jamieson MG (1976) Immunoreactive luteinizing hormone releasing factor in rat pituitary stalk blood: effects of electrical stimulation of the medial preoptic area. J Endocrinol 68:71–87

Fink G, Chiappa SA, Aiyer MS (1976) Priming effect of luteinizing hormone releasing factor elicited by preoptic stimulation and by intravenous infusion and multiple injections of the synthetic peptide. J Endocrinol 69:359–372

Fiske VM (1941) Effect of light on sexual maturation, oestrous cycles, and anterior pituitary of the rat. Endocrinology 29:187–196

Freeman MC, Dupke KC, Croteau CM (1976) Extinction of the estrogen-induced daily signal for LH release in the rat: a role for the proestrus surge of progesterone. Endocrinology 99:223–229

Freeman ME, Neill JD (1972) The pattern of prolactin secretion during pseudopregnancy in the rat: a daily nocturnal surge. Endocrinology 90:1292–1294

Freeman ME, Smith MS, Nazian SJ, Neill JD (1974) Ovarian and hypothalamic control of the daily surges of prolactin secretion during pseudopregnancy. Endocrinology 94:875–882

Gorski RA (1968) Influence of age on the response to perinatal administration of a low dose of androgen. Endocrinology 82:1001–1004

Gorski RA, Barraclough CA (1963) Effects of low dosages of androgen on the differentiation of hypothalamic regulatory control of ovulation in the rat. Endocrinology 73:210–216

Gosden RG, Everett JW, Tyrey L (1976) Luteinizing hormone requirements for ovulation in the pentobarbital-treated proestrous rat. Endocrinology 99:1046–1053

Greep RO (1936) Functional pituitary grafts in rats. Proc Soc Exp Biol Med 34:754–755

Greep RO, Hisaw FL (1938) Pseudopregnancies from electrical stimulation of the cervix in the diestrum. Proc Soc Exp Biol Med 39:359–360

Grieg F, Weisz J (1973) Preovulatory levels of luteinizing hormone, the critical period and ovulation in rats. J Endocrinol 57:235–245

Halász B, Gorski RA (1967) Gonadotrophic hormone secretion in female rats after partial or total interruption of neural afferents to the medial basal hypothalamus. Endocrinology 80:608–622

Halász B, Pupp L (1965) Hormone secretion of the anterior pituitary gland after physical interruption of all nervous pathways to the hypophysiotrophic area. Endocrinology 77:553–562

Halmi NS (1952) Two types of basophils in the rat pituitary: 'thyrotrophs' and 'gonadotrophs' vs. beta and delta cells. Endocrinology 50:140–142

Halmi NS, Krieger D (1983) Immunocytochemistry of ACTH-related peptides in the hypophysis. In: The Anterior Pituitary Gland (AS Bhatnager, ed), Raven Press, New York, pp 1–15

Hammond J Jr (1945) Induced ovulation and heat in anestrous sheep. J Endocrinol 4:169–180

Hammond J Jr, Hammond J, Parkes AS (1942) Hormonal augmentation of fertility in sheep. I. Induction of ovulation, superovulation, and heat in sheep. J Agric Sci 32:308–323

Harlan RE and Gorski RA (1977) Steroid regulation of luteinizing hormone secretion in normal and androgenized rats at different ages. Endocrinology 101:741–749

Harrington FE, Eggert RG, Wilbur RD, Linkenheimer WH (1966) Effect of coitus on chlorpromazine inhibition of ovulation in the rat. Endocrinology 79:1130–1134

Harris GW (1950) Oestrous rhythm, pseudopregnancy and the pituitary stalk in the rat. J Physiol (London) 111:347–360

Harris GW (1972) Humours and hormones. The Sir Henry Dale Lecture for 1971. J Endocrinol 53:ii–xxiii

Harris GW, Jacobsohn D (1952) Functional grafts of the anterior pituitary gland. Proc Roy Soc (London) Ser B 139:263–276

Harris GW, Johnson RT (1950) Regeneration of the hypophysial portal vessels, after section of the hypophysial stalk, in the monkey (Macacus rhesus). Nature (London) 165:819–820

Hartman CG (1944) Some new observations on the vaginal smear of the rat. Yale J Biol Med 17:99–118

Hemmingsen AM, Krarup NB (1937) Rhythmic diurnal variations in the oestrous phenomena of the rat and their susceptibility to light and dark. Kgl Danske Videnskab Selskab, Biol Medd 13(7):1–61

Heuson JC, Waelbroeck-van Gaver C, Legros N (1970) Growth inhibition of mammary carcinoma and endocrine changes produced by 2-Br-α-ergocryptine, a suppressor of lactation and nidation. European J Cancer 6:353–356

Hillarp N-Å (1949) Studies on the localization of hypothalamic centres controlling the gonadotrophic function of the hypophysis. Acta Endocrinol 2:11–23

Hoffman JC (1970) Light and reproduction in the rat: effects of photoperiod length on albino rats from two different breeders. Biol Reprod 2:255–261

Hohlweg W, Chamorro A (1937) Über die luteinisierende Wirkung des Follikelhormons durch Beinflussung der endogenen Hypophysenvorderlappensekretion. Klin Wchnschr 16:196–197

Holck HGO (1942) Dosage of drugs for rats. In: The Rat in Laboratory Investigation (Griffith and Farris, eds) Lippincott, Philadelphia, pp 297–350

Huang HH, Meites J (1975) Reproductive capacity of aging female rats. Neuroendocrinology 17:289–295

Jacobsohn D (1954) Regeneration of hypophysial portal vessels and grafts of anterior pituitary glands in rabbits. Acta Endocrinol (Kbh) 17:187–197

Kalra SP, Sawyer CH (1970) Blockade of copulation-induced ovulation in the rat by anterior hypothalamic deafferentation. Endocrinology 87:1124–1128

Kawakami M, Sawyer CH (1959) Neuroendocrine correlates of changes in brain activity thresholds by sex steroids and pituitary hormones. Endocrinology 65:652–668

Kobayashi F, Hara K, Miyake T (1971) Induction of delayed or advanced ovulation by estrogen in 4-day cyclic rat. Endocrinol Jap 18:389–394

Krey LC, Everett JW (1973) Multiple ovarian response to single estrogen injections early in rat estrous cycles: impaired growth, luteotropic stimulation and advanced ovulation. Endocrinology 93:377–384

Lane CE, Hisaw FL (1934) The follicular apparatus of the ovary of the immature rat and some of the factors that govern it. Anat Rec 60 (Suppl 52):52

Legan SJ, Coon GA, Karsch FJ (1975) Role of estrogen as initiator of daily LH surges in the ovariectomized rat. Endocrinology 96:50–56

Leipheimer RE, Condon TP, Curry JJ (1984) The role of neurotransmitters in mediating copulation-induced ovulation in the rat. J Endocrinol 100:361–365

Lincoln DW and Kelly WA (1972) The influence of urethane on ovulation in the rat. Endocrinology 90:1594–1599

Lipschütz A (1935) Différences prehypophysaires spécifiques du sexe, chez le cobaye Compt rend Soc Biol 118:331–333

Long JA, Evans HM (1922) The oestrous cycle in the rat and its associated phenomena. Memoirs Univ Calif 6:1–111

Martin JE, Tyrey L, Everett JW and Fellows RE (1974a) Variation in responsiveness to synthetic LH-releasing factor (LRF) in proestrous and diestrus-3 rats. Endocrinology 94:556–562

Martin JE, Tyrey L, Everett JW and Fellows RE (1974b) Estrogen and progesterone modulation of the pituitary response to LRF in the cyclic rat. Endocrinology 95:1664–1674

McCann SM, Taleisnik S, Friedman HM (1960) LH-releasing activity in hypothalamic extracts. Proc Soc Exp Biol Med 104:432–434

McCann SM, Ramirez D, Abrams R (1964) Regulation of luteinizing hormone (LH) secretion by a hypothalamic LH-releasing factor. In: Hormonal Steroids, Biochemistry, Pharmacology and Therapeutics: Proc. of the First International Congress of Hormonal Steroids (A Martini and A Pecile, eds) Academic Press, New York pp 251–258

Merchenthaler I, Kovács G, Lavász G, Setalo G (1980) The preopticoinfundicular LH-RH tract of the rat. Brain Res 198:63–74

Moll J, Zeilmaker GH (1966) Ovulatory discharge of gonadotrophins induced by hypothalamic stimulation in castrated male rats bearing a transplanted ovary. Acta Endocrinol 51:281–289

Moore CR, Price D (1932) Gonad hormone functions, and the reciprocal influence between the gonads and hypophysis with its bearing on sex hormone antagonism. Amer J Anat 50:13–72

Murakami N, Takahashi M, Suzuki Y (1978) Conditions for establishment of reflex ovulation in light estrous rats. Endocrinol Japan 25:299–303

Murphy LL and Tyrey L (1986) Induction of luteinizing hormone release by electrochemical stimulation of the medial preoptic area in Δ^9- tetrahydrocannabionol-blocked proestrous rats. Neuroendocrinology 43:471–475

Nequin LG, Schwartz NB (1971) Adrenal participation in the timing of mating and LH release in the cyclic rat. Endocrinology 88:325–331

Nequin LG, Alvarez J, Schwartz NB (1979) Measurement of serum steroid and gonadotropin levels and uterine and ovarian variables throughout 4 day and 5 day estrous cycles in the rat. Biol Reprod 20:659–670

Nichols DC (1969) The timing of the ovulatory surge of gonadotrophin induced by estrogen in cyclic and pseudopregnant rats. MA Thesis, Duke University

Nikitovitch-Winer M (1957) Humoral influence of the hypothalamus on gonadotrophin secretions. PhD Dissertation, Duke University

Nikitovitch-Winer MB (1965) Effect of hypophysial stalk transection on luteotropic hormone secretion in the rat. Endocrinology 77:658–666

Nikitovitch-Winer M, Everett JW (1958a) Comparative study of luteotropin secretion by hypophysial autotransplants in the rat. Effects of site and stages of the estrus cycle. Endocrinology 62:522–532

Nikitovitch-Winer M, Everett JW (1958b) Functional restitution of pituitary grafts re-transplanted from kidney to median eminence. Endocrinology 63:916–930

Nikitovitch-Winer M, Everett JW (1959) Histocytologic changes in grafts of rat pituitary on the kidney and upon re-transplantation under the diencephalon. Endocrinology 65:357–368

Nir I, Ayalon D, Tsafiri A, Cordova T, Lindner HA (1973) Suppression of cyclic surge of luteinizing hormone secretion and ovulation in the rat by Δ^9-tetrahydrocannabinol. Nature (London) 243:470–471

Phillips WA (1937) The inhibition of estrous cycles in the albino rat by progesterone. Amer J Physiol 119:623–626

Pickering AJMC, Fink G (1979) Priming effect of luteinizing hormone releasing factor *in vitro*: role of protein synthesis, contractile elements, Ca^{++} and cyclic AMP. J Endocrinol 81:223–234

Purves HD, Griesback WE (1951) The site of thyrotrophin and gonadotrophin production in the rat pituitary studied by McManus-Hotchkiss staining for glycoprotein. Endocrinology 49:244–264

Quilligan EJ, Rothchild I (1960) The corpus luteum-pituitary relationship: the luteotrophic activity of homotransplanted pituitaries in intact rats. Endocrinology 67:48–53
Quinn DL (1966) Luteinizing hormone release following preoptic stimulation in the male rat. Nature (London) 209:891–892
Quinn DL, Everett JW (1967) Delayed pseudopregnancy induced by selective hypothalamic stimulation. Endocrinology 80:155–162
Redmond WC(1968) Ovulatory response to brain stimulation or exogenous luteinizing "of" hormone in progesterone-treated rats. Endocrinology 83:1013–1022
Reid SA, Sypert GW (1980) Acute $FeCl_3$-induced epileptogenic foci in cats: electrophysiological analyses. Brain Res 188:531–542
Roser S, Block RB (1971) Étude comparative des variations de la progestérone plasmatique ovarienne au cours de cycles respectivement de 4 et de 5 jours chez la ratte. Compt rend Soc Biol 165:1995–1998
Rothchild I (1981) The regulation of the mammalian corpus luteum. Rec Prog Horm Res 37:183–283
Sanchez-Criado J, Rothchild I (1986) The relation between the effects of hysterectomy, decidual tissue, prolactin, or luteinizing hormone (LH) and the ability of indomethacin to prevent luteolysis in rats bearing LH-dependent corpora lutea. Endocrinology 119:1750–1756
Sawyer CH, Critchlow BV, Barraclough CA (1955) Mechanism of blockade of pituitary activation in the rat by morphine, atropine and barbiturates. Endocrinology 57:345–354
Sawyer CH, Everett JW (1959) Stimulatory and inhibitory effects of progesterone on the release of pituitary ovulating hormone in the rabbit. Endocrinology 65:644–651
Sawyer CH, Everett JW, Markee JE (1949) A neural factor in the mechanism by which estrogen induces the release of luteinizing hormone in the rat. Endocrinology 44:218–233
Sawyer CH, Markee JE, Hollinshead WH (1947) Inhibition of ovulation in the rabbit by the adrenergic-blocking agent Dibenamine. Endocrinology 41:395–402
Sawyer CH, Markee JE, Townsend BF (1949) Cholinergic and adrenergic components in the neurohumoral control of the release of LH in the rabbit. Endocrinology 44:18–37
Schuiling GA, De Koning J, Zürcher AF (1976) On differences between the preovulatory luteinizing hormone surges of 4- and 5-day cyclic rats. J Endocrinol 70:373–378
Schwartz NB (1969) A model for the regulation of ovulation in the rat. Recent Progr Hormone Res 25:1–43
Seegal RF, Goldman BD (1975) Effects of photoperiod on cyclicity and serum gonadotropins in the Syrian hamster. Biol Reprod 12:223–231
Sherwood NM, Chiappa SA, Sarkar DK, Fink G (1980) Gonadotropin-releasing hormone (GnRH) in pituitary stalk blood from proestrous rats: effects of anesthetics and relationship between stored and released GnRH and luteinizing hormone. Endocrinology 107:1410–1417
Smith PE (1961) Postponed homotransplants of the hypophysis into the region of the median eminence in hypophysectomized male rats. Endocrinology 68:130–143
Smith PE (1963) Postponed pituitary homotransplants into the region of the hypophysial portal circulation in hypophysectomized female rats. Endocrinology 73:793–806
Smith PE, White WE (1931) The effect of hypophysectomy on ovulation and corpus luteum formation in the rabbit. J Amer Med Assn 97:1861–1863
Stockard CR, Papanicolaou GN (1917) The existence of a typical oestrus cycle in the guinea pig – with a study of its histological and physiological changes. Amer J Anat 22:225–283
Swanson H, van der Werff ten Bosch JJ (1964) The "early androgen" syndrome, its development and the response to hemispaying. Acta Endocrinol 45:1–12
Szentágothai J, Flerkó B, Mess B, Halász B (1968) Hypothalamic control of the anterior pituitary. Akadémiai Kiadó, Budapest

Takeo Y (1984) Influence of continuous illumination on estrous cycle of rats: time course of changes in levels of gonadotropins and ovarian steroids until occurrence of persistent estrus. Neuroendocrinology 39:97–104

Taleisnik S, Sherwood MRC, Raisman G (1979) Dissociation of spontaneous and mating induced ovulation by frontal hypothalamic deafferentations in the rat. Brain Res 169:155–162

Taubenhaus M, Soskin S (1941) Release of luteinizing hormone from the anterior hypophysis by an acetylcholine-like substance from the hypothalamic region. Endocrinology 29:958–964

Tejasen T, Everett JW (1967) Surgical analysis of the preoptico-tuberal pathway controlling ovulatory release of gonadotropins in the rat. Endocrinology 81:1387–1396

Terasawa E, Bridson WE, Weishaar DJ, Rubens LV (1980) Influence of ovarian steroids on pituitary sensitivity to luteinizing hormone-releasing hormone in the ovariectomized guinea pig. Endocrinology 106:425–429

Terasawa E, Kawakami M, Sawyer CH (1969) Induction of ovulation by electrochemical stimulation in androgenized and spontaneously constant-estrous rats. Proc Soc Exper Biol Med 132:497–555

Terasawa E, Rodriguez JS, Bridson WE, Wiegand SJ (1979) Factors influencing the positive feedback action of estrogen upon the luteinizing hormone surge in the ovariectomized guinea pig. Endocrinology 104:680–686

Terasawa E, King MK, Wiegand SJ, Bridson WE, Goy RW (1979) Barbiturate anesthesia blocks the positive feedback effect of progesterone, but not of estrogen, on luteinizing hormone release in ovariectomized guinea pigs. Endocrinology 104:687–692

Tobin CF (1942) Effects of lactogen on normal and adrenalectomized rats. Endocrinology 31:197–200

Turgeon J, Barraclough CA (1973) Temporal patterns of LH release following graded preoptic electrochemical stimulation in proestrous rats. Endocrinology 92:755–761

Tyrer NM, Bell EM (1974) The intensification of cobalt-filled neurone profiles using a modification of Timm's sulphide-silver method. Brain Res 73:151–155

van der Schoot P (1976) Changing pro-oestrous surges of luteinizing hormone in ageing 5-day cyclic rats. J. Endocrinol. 69:287–288

van der Schoot P, Lincoln DW, Clark JS (1978) Activation of hypothalamic neuronal activity by electrolytic deposition of iron into the preoptic area. J. Endocrinol 79:107–120

van der Schoot P, Uilenbroek JTJ (1983) Reduction of 5-day cycle length of female rats by treatment with bromocriptine. J Endocrinol 97:83–89

Velasco ME, Rothchild I (1973) Factors influencing the secretion of luteinizing hormone and ovulation in response to electrochemical stimulation of the preoptic area in rats. J Endocrinol 58:163–176

Waterman AJ (1943) Studies of normal development of the New Zealand White strain of rabbit. Amer J Anat 73:473–515

Weick RF, Davidson JM (1970) Localization of the stimulatory feedback effect of estrogen on ovulation in the rat. Endocrinology 87:693–700

Westman A (1942) Der Einfluss des Hypophysen-Zwischenhirnsystems auf die Sexualfunktionen. Schweiz Med Wchnschr 72:113–116

Westman A, Jacobsohn D (1936) Über Ovarialveränderungen beim Kaninchen nach Hyophysktomie. Acta Obst Gynecol Scand 16:483–508

Westman A, Jacobsohn D (1937) Experimentelle Untersuchungen über die Bedeutung des Hypophysen-Zwischenhirnsystems fur die Produktion gonadotroper Hormone des Hypophysenvorderlappens. Acta Obst Gynec Scand 17:235–265

Westman A, Jacobsohn D (1938a) Endokrinologische Untersuchungen an Ratten mit durchtrennten Hypophysenstiel. I. Hypophysenveränderungen nach Kastration und nach Oestrinbehandlungen. Acta Obstetr Gynec Scand 18:99–108

Westman A, Jacobsohn D (1938b) Endokrinologische Untersuchungen an Ratten mit durchtrennten Hypophysenstiel. III. Über die luteinizierende Wirkung des Follikelhormons. Acta Obst Gynec Scand 18:115–123

Westman A, Jacobsohn D (1938c) Endokrinologische Untersuchungen an Ratten mit durchtrennten Hypophysenstiel. VI. Produktion und Abgabe der gonadotropen Hormone. Acta Path Microbiol Scand 15:445–453

Willmore LJ, Hurd RW, Sypert GW (1978a) Epileptiform activity initiated by pial iontophoresis of ferrous and ferric chloride on rat cerebral cortex. Brain Res 152:406–410

Willmore LJ, Sypert GW, Munson JB (1978b) Recurrent seizures induced by cortical iron injection: a model of posttraumatic epilepsy. Ann Neurol 4:329–336

Willmore LJ, Hiramatsu M, Kochi H, Mori A (1983) Formation of superoxide radicals after $FeCl_3$ injection into rat isocortex. Brain Res 277:393–396

Wise PM, Camp-Grossman P, Barraclough CA (1981) Effects of estradiol and progesterone on plasma gonadotropins, prolactin, and LHRH in specific brain areas of ovariectomized rats. Biol Reprod 24:820–830

Wolfe JM, Bryan WR, Wright AW Histologic observations on the anterior pituitaries of old rats with particular reference to the spontaneous appearance of pituitary adenomata. Amer J Cancer 34:352–372

Wolfe JM, Burack E, Wright AW (1940) The estrous cycle and associated phenomena in a strain of rats characterized by a high incidence of mammary tumors together with observations on the effects of advancing age on these phenomena. Amer J Cancer 38:383–398

Wuttke W, Meites J (1973) Effects of electrochemical stimulation of medial preoptic area on prolactin and luteinizing hormone release in old female rats. Pflügers Arch 341:1–6

Ying S-Y, Greep RO (1972) Effect of a single injection of estradiol benzoate (EB) on ovulation and reproductive function in 4-day cyclic rats. Proc Soc Exp Biol Med 139:741–744

Ying S-Y, Meyer RK (1969) Effect of coitus on barbiturate-blocked ovulation in immature rats. Fertil Steril 20:772–778

Zarrow MX, Clark JM (1968) Ovulation following vaginal stimulation in a spontaneous ovulator and its implications. J Endocrinol. 40:343–352

Zeilmaker GH (1965) Normal and delayed pseudopregnancy in the rat. Acta Endocrinol (Kbh) 49:558–566

Zeilmaker GH (1966) The biphasic effect of progesterone on ovulation in the rat. Acta Endocrinol 51:461–468

Zuckerman S (1938) Cyclical fluctuations in the sensitivity of the rat to oestrogenic stimulation. J Physiol 92:12P

Acknowledgments

Support for the investigations in the author's laboratory before 1960 came from occasional small grants from the Duke University Research Council, supplementing internal funds from the Department of Anatomy. The studies by Dr. Nikitovitch-Winer published in 1957–1959 were also supported in part by a grant from the Committee for Research in Problems of Sex, National Academy of Sciences-National Research Council. From 1960 onward the principal support was from a series of grants from the National Science Foundation.

Special mention must be made of the excellent photomicrography by Carl Bishop of the Department of Pathology for the illustrations appearing in our publications before 1960. All radioimmunoassays of pituitary hormones were carried out under the supervision of Dr. Lee Tyrey. I am indebted to Dr. Tyrey and to Dr. Charles H. Sawyer for their critical reviews and helpful suggestions in the preparation of this monograph, as well as for their active participation in much of the research. Important contributions were made by several students, fellows and visiting investigators, much of whose research is cited in the text; specifically: Judith E. Beach, Joseph P. Bunn, Judith Furman, Roger G. Gosden, James W. Holsinger, Lewis C. Krey, Doris C. Nichols, Miroslava Nikitovitch-Winer, David L. Quinn, Max Radford, William C. Redmond, Linda Smith, Tejatat Tejasen, and Gerard H. Zeilmaker.

Most of the illustrations and tables in the monograph are reproduced from our publications in *Endocrinology*. I am also indebted for the use of certain items from the *American Journal of Anatomy* (Figs. 11 and 12) and the *Journal of Endocrinology* (Tables 8–10). Permission for use of other material was generously granted by the *Academic Press* (Figs. 34–37); *Centre National de la Recherche Scientifique, Paris* (Figs. 2–6, 28 and 29); *Gordon and Breach*, N.Y. (Fig. 40); *S. Karger*, Basel/N.Y. (Figs. 40, 46 and Table 18); *Pergamon Press, New York, Oxford* (Fig. 33).

The careful assistance by Ms. Dana Hall in preparing this manuscript is greatly appreciated.

Index

ACTH, 61, 99, 101, 105
Action potentials, effect of iron deposits, 61–62
Activation period, 56
Age effects, 9–10, 14–15
Amygdala, 57
Anterior hypothalamic area (AHA), 52, 58, 63
Anterior pituitary (AP) grafts, 93–106
AP graft, blood supply, 100
AP transplants near MEm, 100–101, 104–106
Arcuate nucleus (ARC), 75–82, 89–90

Cervical stimulation, 1, 39, 92–93, 106
Circadian rhythm, LH surge, 51–55, 63
Circadian rhythm, estrous behavior, 5, 38, 52
Continuous light, 14–15, 20, 57
Copper chloride, 58
Copulation, 5, 106
Corpora lutea, strain differences, 24–27
Corpus luteum cholesterol, 24–35, 39–41
  Enlargement by estrogen, 95
  Function in estrous cycle, 24–35, 92
  Regression, 92
  Responsiveness, 95–97
Critical period: LH surge, ovulation, 39, 46–47, 50–56

Decidual reaction, 1, 93, 99, 106, 108
Delayed pseudopregnancy, 106–114
  Follicle atresia, 109–110
  Genital stimulation, 106–107, 110–112
  Hypothalamic stimulation, 107–110, 112
  Prolactin surges, 109–114
  Stimulus degree/duration of effect, 108–109, 114

Diagonal band, 65
Diestrus-3 vs. Proestrus, 68–71
Dorsomedial/ventromedial hypothalamus (DMH/VMH), 107–108, 112–113
Dose-response, 66, 68–73, 75–82
  Brain stimulation/LH release; ovulation, 66, 68–73, 75–82, 84
  LH/ovulation, 68–69, 71, 77–80
  LHRH/LH release, 69

Electrical stimulation (ELS), 62, 73–90
  Cycling rat, pharmacologic block, 62, 73–87
  MPOA, ARC/ME sites compared, 75–82, 89–90
  Periodic pulse trains, 84–85
  Persistent estrus, 87–90
  Stimulus duration/response, 75, 80–82
Electrical parameters, 58–59, 62, 65, 73
Electrochemical stimulation (ECS), 57–66, 68–73, 83–84, 87–90
  Cycling rat, pharmacologic block, 57–62, 68–73, 83–84
  Iron deposits, 60–62
  Lesion characteristics, 59, 62
  Male, castrate with ovary, 88
  Persistent estrus, 87–90
  Pseudopregnancy, 90
Electrode design, 58–59, 62
Electrode sites, 57–58, 63–65, 74–75, 78, 85, 87, 113
Ergotamines, 32–33, 114
Estrogen, 33–35, 39–41, 53, 56, 92
  Advance of ovulation time, 35, 39
  AP response to LHRH enhanced, 69
  Corpus luteum cholesterol, 33–35
  Ovulation: pregnancy, pseudopregnancy, 33–35

Pharmacologic blockade, 39–41
Strain, species differences, 56
Estrous cycle, 1–7, 32–33, 38, 100–101, 104–106
Duration, 1, 7
Normal features, 1–7
Ovulation time, 5, 38
Renewed cycles: persistent estrus, 32–33

Graafian follicle,
Atresia, 53, 55, 109–110
Growth, 53, 55
Partial effects of LH, 38, 42–43, 48–49, 56, 71, 77, 87
Growth hormone, 105

Hypophysectomy, 34, 47–49, 60, 71–72, 78–79, 93, 99–100, 105
Hypothalamic surgery, (cuts), 57, 66–67

Interstitial tissue, 27, 94–95, 100–101
Iron, irritant effects, 61–62

Lateral preoptic area (LPOA), 65, 73–75
Lesions, ECS: iron deposits, 58–59, 62, 66
Lesions, radiofrequency, 66–67, 69
Lesions, tissue damage, 57, 62, 66–69
Luteinizing hormone (LH): ovulatory surge, 3, 5, 38, 47, 50, 66–69, 71–73, 77–86, 89–90
LH release time, 69, 71–72
Luteinizing hormone-releasing hormone (LHRH), 50, 61, 69, 87

Medial basal tuber (MBT), 52, 63, 66
Medial forebrain bundle (MFB), 65
Medial parolfactory area, 65
Medial preoptic area (MPOA), 58, 60, 64–66, 70–72, 83–88
Medial preoptic nucleus (MPN), 65
Median eminence (ME, MEm), 52

Nephrectomy, 79
Nucleus accumbens, 65

Ovariectomy, 53
Ovulation blocking agents, 38–44, 46–48, 51–54, 56–57, 82–84, 86–87, 106–114
Atropine, 38–44, 46, 51–52, 56, 83–84, 86
Dibenamine, 38–41, 43–44, 46
Pentobarbital, 43, 47, 52–54, 57, 83–84, 86, 106–114
Other, 47–48, 82–84, 87
Ovulation, luteinization, 5, 18–19, 22–23, 33–35, 37–38, 46, 90
Advanced by estrogen, 35
Advanced by progesterone, 22–23
Induced in persistent estrus, 18–19, 37
Induced in pregnancy, pseudopregnancy, 33–35, 90
Time in estrous cycle, 5, 38, 46
Ovulation reflex, 47, 57, 92, 106–107
Ovulatory quota of LH, 39, 49–50, 69, 72, 79–80

Persistent estrus (PE), 5–11, 14–15, 18–20, 32–33, 57, 65, 87–88
AHA lesions, 65
Effects of age, 9–10, 14–15
In continuous light (LLPE), 14–15, 20, 57
Androgenized rats (TPPE), 10, 87–88
Spontaneous (SPE), 6, 7, 9, 11, 18–19, 32–33
Strain differences, 8-9, 14–15
Photoperiodicity, 11–14, 16, 51, 56
Pituitary portal vessels, 50, 87, 89
Potential activation period, 56
Progesterone effects, 17–24, 33–35, 37, 43–45, 52–56, 69
AP response to LHRH enhanced, 69
Biphasic action, 17, 23–24, 53, 56
Cycling renewed: LLPE, SPE, 18–20, 27
Interaction with estrogen, 20–21, 33–35, 37, 52, 69
LH release, ovulation, 18–24, 42–45
Pharmacologic blockade, 43–44
Prolactin (PRL) effects, 27–33, 92
Cycling renewed: SPE, 27–33
Luteotropic action, 27, 92
Prolactin secretion, 27–29, 31–33, 53, 93–106
AP grafts, 93–106
Ergotamine blockade, 31–33
Estrous cycle, 27–29

Inhibition by hypothalamus, 93, 106
Old rats, 32
Stalk-section, 92–93
Prolonged diestrus, 11–13
Pseudocyclicity, 7
Pseudopregnancy, pregnancy, 1, 33–35, 39–41, 90, 92–93, 98–99
AP grafts, 93–98
Damage to MBT, stalk, AP, 95, 98–99
Genital stimulation, 1, 39, 92–93
Ovulation by brain stimulation, 90
Ovulation, luteinization by estrogen, 33–35, 39–41
Stalk-section, 92, 98–99

Secondary corpora lutea, 97
Septal-preoptic tuberal system (SPTS), 63–66, 72, 80, 82, 87
Septum pellucidum, 57, 61, 65
Strain differences, 8, 9, 24, 34–35, 37–39, 72–73, 75–76, 87–88
Stress, 56
Stria terminalis, bed nucleus, 65
Suprachiasmatic region, nucleus, 65–66

Tissue damage, 57, 62, 66–69
TSH, 101, 105

MIX
Papier aus verantwortungsvollen Quellen
Paper from responsible sources
FSC® C105338

If you have any concerns about our products,
you can contact us on
ProductSafety@springernature.com

In case Publisher is established outside the EU,
the EU authorized representative is:
Springer Nature Customer Service Center GmbH
Europaplatz 3, 69115 Heidelberg, Germany

Printed by Libri Plureos GmbH
in Hamburg, Germany